U0334047

酒店配色与细部解析

欧式奢华

度本图书 编著

中国林业出版社

西方古典主义风格以其华丽的设计、浓烈的色彩、精美的造型和考究的软装效果自始至终在建筑和室内美学中占有重要的地位。古典风格里设计师对美的追求跨越了古希腊、古罗马、哥特、拜占庭、文艺复兴、巴洛克、洛可可、浪漫主义、折中主义等时空与流派的界限，其表现出的文化传承和精神内涵也历久弥新地震撼着我们的心灵。

不同的语言，表达着不同的思想，流露出不同的情感；不同的建筑，承载着不同的文化，体现着不同的信念。蕴含在一栋建筑之中的美，有着不一样的表现形式，现代人用新的手法和技术去还原古典建筑里的雍容华贵和浪漫情怀，也是在工业化的时代对永恒的美学标准表达着尊崇与敬意。

本书在全球范围内精选了二十多例融入西方古典风格的当代酒店建筑及室内作品，从配色、材质、光线等角度，通过对设计细节的剖析，向读者展示了现代设计理念引领下的新古典美学所呈现出的独特艺术魅力。

Dopress Books

目录

Hotel Colors & Details

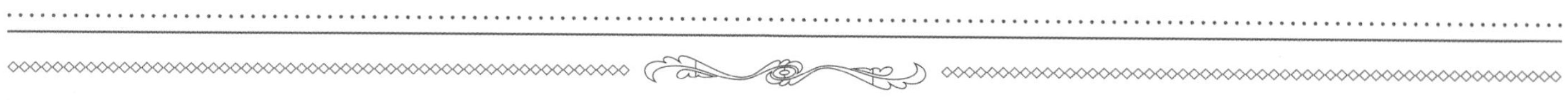

目录

Hotel Colors & Details

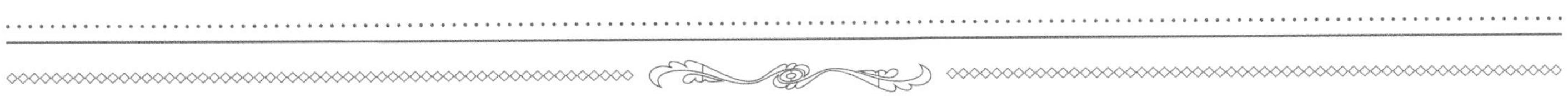

Four Seasons Hotel des Bergues Geneva

德贝尔格四季酒店（瑞士）

新古典风格的古老酒店“贝尔格”始建于1843年，历经改造后在2005年成为四季酒店集团的一员。法国设计师Pierre-Yves Rochon为酒店设计了标志。象征日内瓦湖水的蓝色和绿色，衬托了Serge Marzetta的花卉摆设。酒店曾作为早期国家联盟的场所，现在已成为商务人士钟爱的地点，同时也是那些被市内时尚店铺、历史古迹、博物馆和安静场所及酒店餐馆吸引的人们的圣地。

五个皇家套房内卧室的举架很高，高大的落地窗和优雅的地板，都带有宫殿般的奢华。家具全部为法国制造，散发着优雅的凡尔赛气息。华丽的丝绸、天鹅绒和提花布料都来自巴黎高级纺织品店。极其精致复杂的建筑细节包括泥灰装饰线条、雕刻、涡卷纹饰全部为现场制作，手工喷涂，以保留历史古韵。

酒店在2013年迎来了第178周年店庆。自1834年首次营业，贝尔格四季酒店在革新后的道路上蒸蒸日上。酒店最新开放了两个高级套房，全部为法国著名设计师Pierre-Yves Rochon设计。这两个套房在欧洲也是顶级奢华。酒店对于那些希望找到“家”的感觉的住客来说是首选。每层楼内的私人助手可为日内瓦四季酒店贝尔格的每个套房，提供最方便的服务。在每位客人的入住期间，经过严格培训的人员会提供细致入微的服务。其中包括酒店私人司机去机场接住客，在正门将客人用豪车送达套房。对客人的个人喜好如鲜花、音乐、矿泉水和下午茶都掌握得精准无误，确保让住客体验到四季酒店的人性化服务。

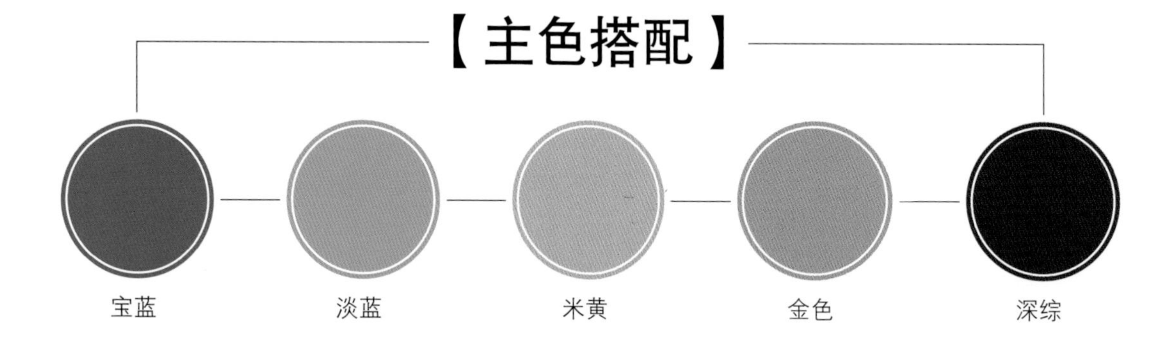

• Area / 占地面积: 16, 000 m^2 • Date of Completion / 竣工时间: 2013 • Interior Design / 室内设计: Pierre-Yves Rochon • Photography / 摄影: Waite, Richard •

LES BERGUES

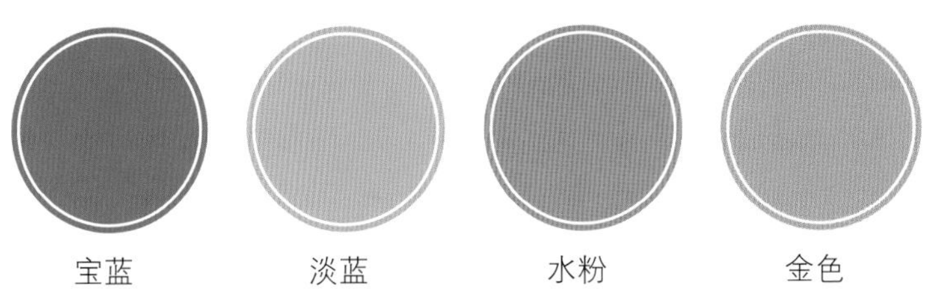

【主题色彩含义】：湖水、浪漫、温馨

湖水般透明清澈的蓝色系点缀水粉色的花朵，让人倍感温馨浪漫。明亮的金色更为房间添上一抹午后斜阳般温暖的光辉。

LA TERRE VUE DU CIEL

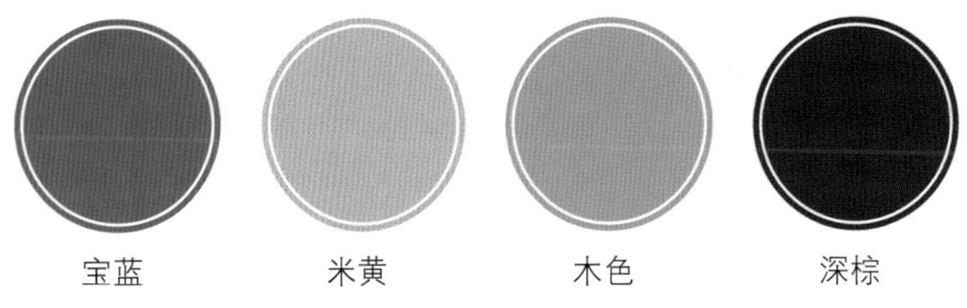

宝蓝　　米黄　　木色　　深棕

【主题色彩含义】：柔和、温暖、生动

从米黄色调的背景、沙发，到天然木色的地板，再到深综色沙发背景墙、写字台和沙发靠垫，咖色系的大面积运用使整个房间都笼罩着一层暖洋洋的光辉。蓝色的地毯、鸡尾酒和靓丽的花朵巧妙地点缀其中，营造出柔和却不乏生动的空间。

紫色

米黄

木色

深棕

【主题色彩含义】：高贵、典雅

在室内设计配色中，运用互补色和对比色有着异曲同工之妙，运用得当都可以起到起到赏心悦目的视觉效果。色相反差极大的“黄”和“紫”，通过饱和度与色彩比例的调和，搭配在一起十分协调优美，也使整个房间看上去更加高贵典雅。

1	3
2	

【细部解析】

1、绒面靠垫、粗纹提花沙发和皮草床盖构成了材质的对比，使单色调的卧室看上去并不单调，反而是充满变化和质感。

2、白绿相间的花束在平淡柔和的咖色调中空间起到画龙点睛的作用。

3、卫浴间中陈列的鲜花和装饰器皿营造出不凡的格调。

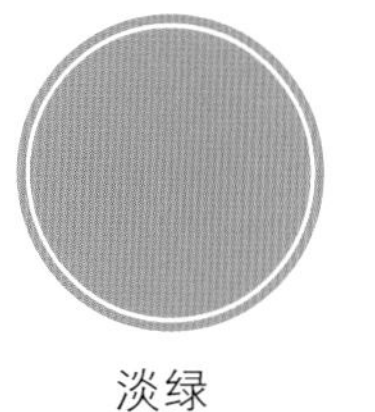
淡绿

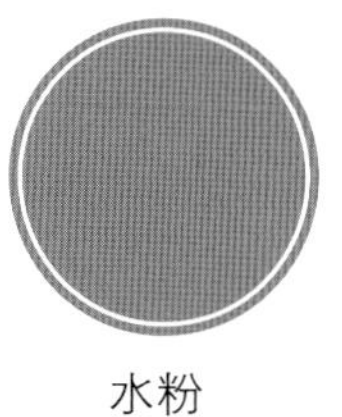
水粉

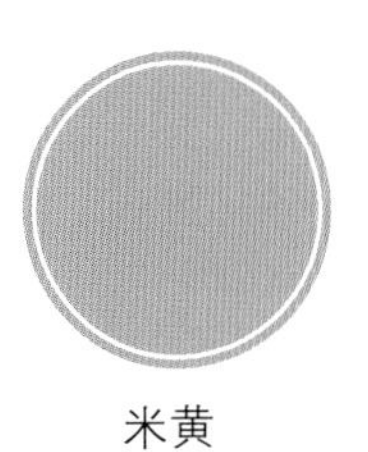
米黄

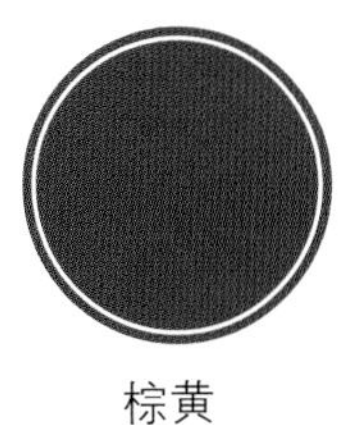
棕黄

【主题色彩含义】：清新、明媚、柔美、雅致

古典主义设计中不可或缺的米色和棕色，原本给人平稳、宁静、古雅的感觉。以其为背景的空间中点缀少量鲜嫩的水粉色和淡绿色，顿时生机盎然，令人耳目一新。

【细部解析】

套房中的家具全部为法国制造，散发着优雅的凡尔赛气息。华丽的丝绸、天鹅绒和提花布料都选自巴黎高级纺织品店。单色调的卧室采用不同面料、花纹的布艺装饰组合，突出的非凡质感，暗示着这里是安静、尊贵、典雅的休息之所。

Macalister Mansion

麦卡利斯特酒店（马来西亚）

麦卡利斯特酒店的设计理念是打造一个体验奇异和精致生活方式的场所，客人们在这里会体验到不同于家的感受。酒店的前身是一栋有着百年历史的豪宅，现在被改造成为一座具有丰富体验感的设计型酒店。

酒店分为6个部分：5个餐厅区和1个客房区。每个部分都展现出酒店无处不在的豪华感 —— 公共餐厅、小客房、书房、酒吧、会客厅、草坪区和八个套房。麦卡利斯特楼的命名源于以前的主人英国总督上校诺曼·麦卡里斯特。人们会发现许多处特意安排的艺术品贯穿整个空间和地点，这些都是参照历史而摆设的。

该设计赢得了2013年餐饮设计与家具展览的亚洲奢华项目组金奖（最佳奢华酒店设计）、2013亚太不动产最佳小型酒店奖（国际不动产奖，英国），同时入围2013年世界最佳新建酒店奖。世界最佳新建酒店奖的评委Juliet Kinsman评价道：Ministry of Design设计的麦卡利斯特酒店在评委会中引起了热议。它以独到的设计手法整合空间，为仅有的八套客房提供了难以置信的设计灵感。该设计营造出了优雅奢华的感觉。最佳奢华酒店奖是为那些成功打造品牌和采用独特设计理念的顶级奢华酒店而设置。赞赏麦卡利斯特酒店设计的评委评价：这是独特的位置和建筑。尤其是定制的艺术品更是锦上添花，彰显了其独特的品位。

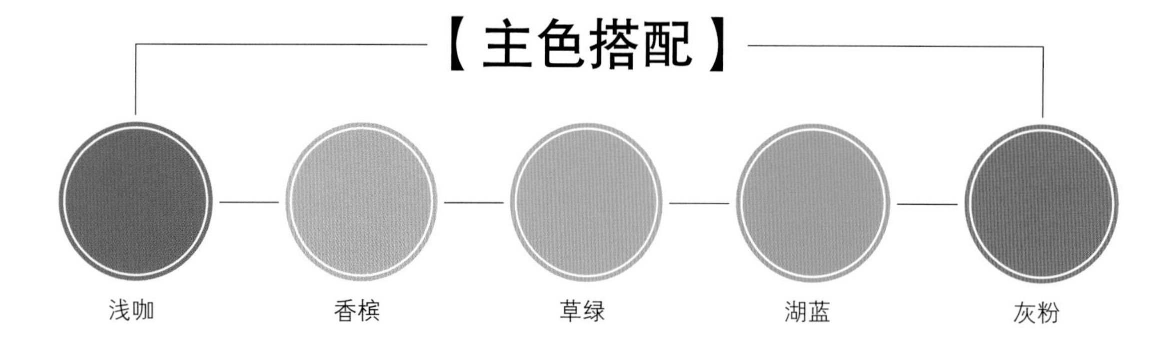

Area / 占地面积: 1,700 m^2 • **Date of Completion / 竣工时间:** 2012 • **Interior Design / 室内设计:** Modonline, Ministry of Design • **Photography / 摄影:** Modonline • **Client / 客户:** Macalister Mansion •

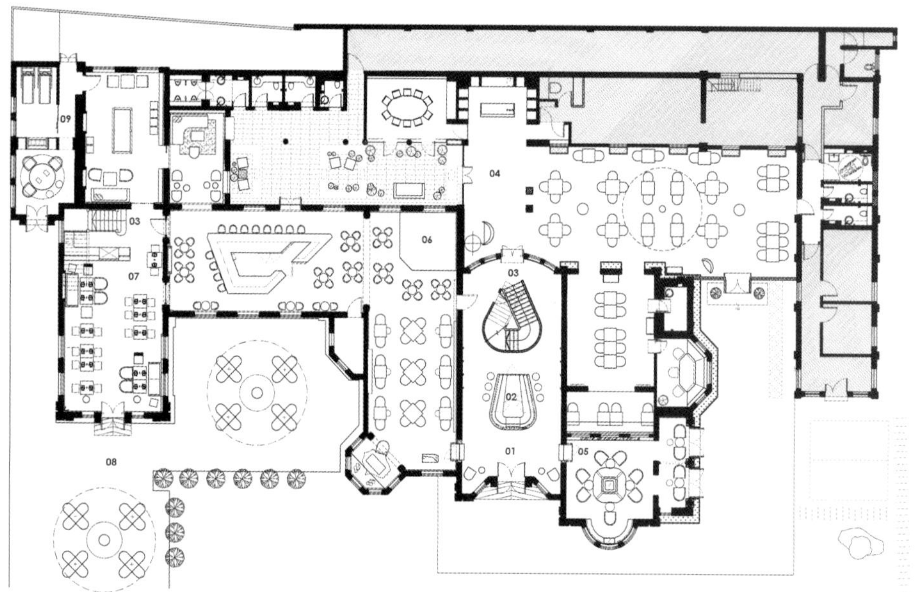

01 DROP OFF
02 RECEPTION
03 TO GUESTROOM
04 DINING ROOM
05 THE DEN
06 BAGAN BAR
07 LIVING ROOM
08 LIVING ROOM (ALFRESCO)
09 HOTEL GUEST LOUNGE

GF LAYOUT PLAN

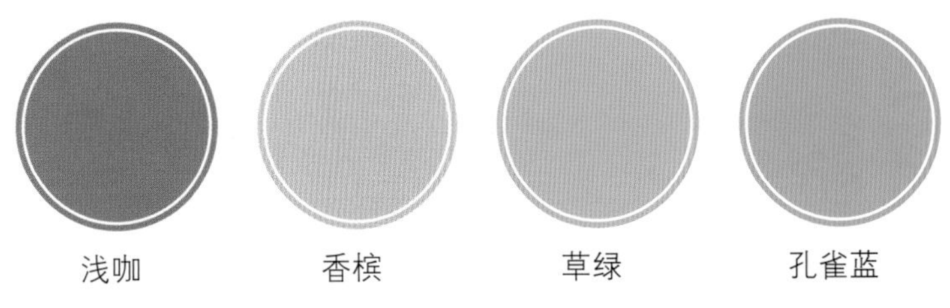

浅咖 香槟 草绿 孔雀蓝

1 | 3
2

【细部解析】

1、酒店前台背景墙是简单的多媒体屏幕。

2、正门把手采用实木材质，上面刻有该酒店的标识。

3、木质台面与白色裸砖墙面和前台立面的反光质地形成对比。

孔雀蓝　　灰蓝　　灰粉

【主题色彩含义】：童话、纯净、唯美

在大面积洁白的背景衬托下，孔雀蓝、灰蓝与灰粉的搭配为餐厅营造出纯净、梦幻、维美的氛围。白色的叶子、树干和淡色的小鹿陈列其中，更使空间充满了童话般美好的意境。

1 | 3
2

【细部解析】

1、马赛克地面的花色衬托出整个房间的优雅高贵的基调。

2、The Den（吸雪茄喝威士忌的场所）入口门把手简约大方。

3、房间中心位置的枝型吊灯为设计增添了一丝古典气息。

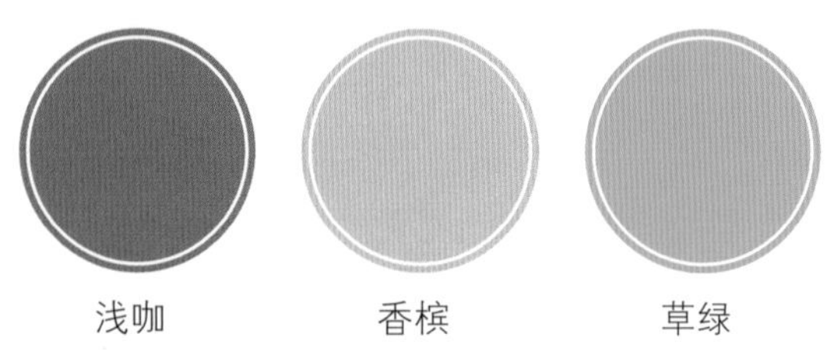

【主题色彩含义】：田园、清新、美味

咖色与绿色的搭配堪称是最为经典的配色方案之一，尤其是运用到餐厅、厨房这类就餐空间。同时结合白色、金色或深综色，更能很好地营造出充满田园生态味道的就餐环境。

Come in

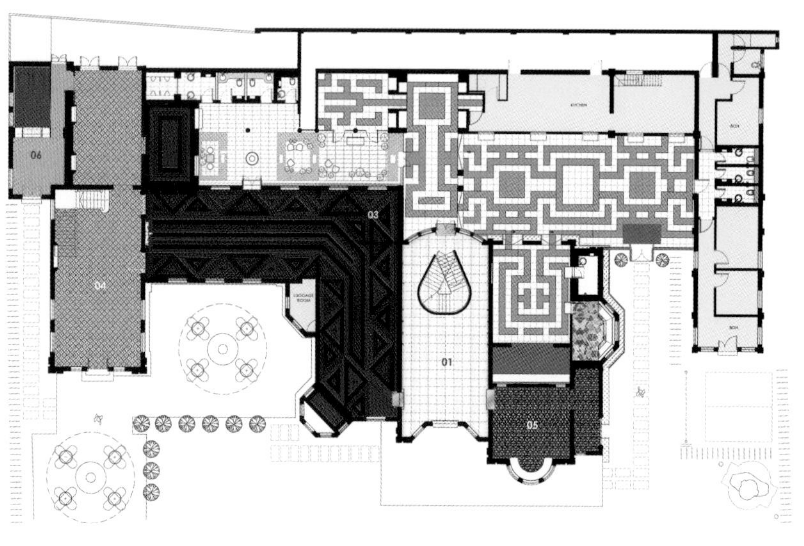

01 RECEPTION
02 DINING ROOM
03 BAGAN BAR
04 LIVING ROOM
05 THE DEN
06 HOTELGUEST LOUNGE

MACALISTER MANSION　　GF ZONING PLAN

大红　　黑色

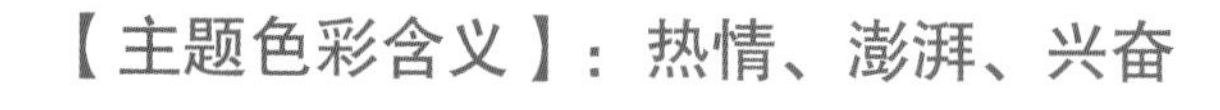

【主题色彩含义】：热情、澎湃、兴奋

红色运用在酒店、餐厅和零售空间中屡见不鲜。正如米歇尔·帕斯图罗（Michel Pastoureau）所说："谈论'红色'几乎是多余的。红色就是纯粹、经典，所有颜色之最。"

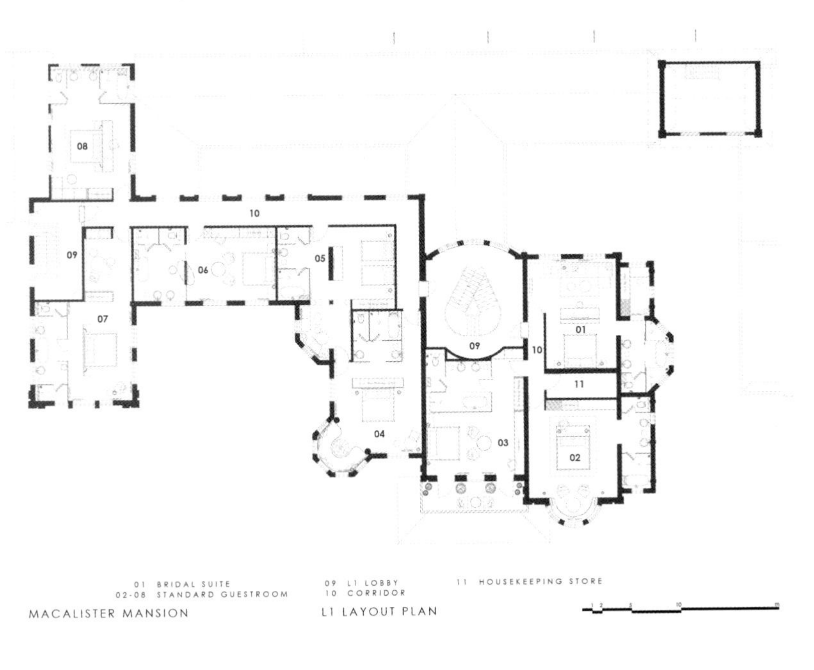

	2
1	3

【细部解析】

1、吊灯、壁灯和地灯的光线把原本单调的空间塑造得层次分明。

2、墙壁上的线描不难让人联想到酒店悠久的历史背景。

3、定制的艺术品和家具让整体设计显得更加别具一格。

【细部解析】

每个房间中都能看到设计师精心布置的软装设计。或者是造型独特的家具，或者是耐人寻味的墙饰及壁画，或者是图案精美的瓷砖、地砖，每一件都给人带来耳目一新的感觉。

InterContinental London Park Lane
洲际伦敦公园弄度假酒店（英国）

HBA设计团队以伊丽莎白二世女王为灵感来源，将伦敦帕克巷洲际酒店的皇家套房打造成拥有永恒魅力的经典。套房在2012年对外营业，正值女王钻石婚年，设计师们被酒店的尊贵位置所感染，在得知女王陛下的时尚品味，尤其是在观察塞西尔·比顿为年轻时的女王拍摄的照片后，HBA设计了该套房。新的设计也表达了对女王作为和蔼的家庭一员和世界名人的敬意。

本设计将装饰艺术以丰富的现代的细节融合入皇家宫殿的设计元素。创造性地将材质的光泽质感融入到室内空间的设计中，以此提升空间感和明亮度。起居室中，古风镜面边框加上天花板吊顶的角线营造出更高的空间感，咖啡边桌、茶几、灯饰都有光亮的金色装饰，簇绒皮革长椅同样由光亮底座支撑，看起来十分精致。就餐区内有为套房特制的精美鸡尾酒推车，还有一个全尺寸灰色玻璃瓦尔（Bolivar）饰面支架，上面摆有电视。由于侧拉门的设计，使客人在起居室和就餐区都可以观看。

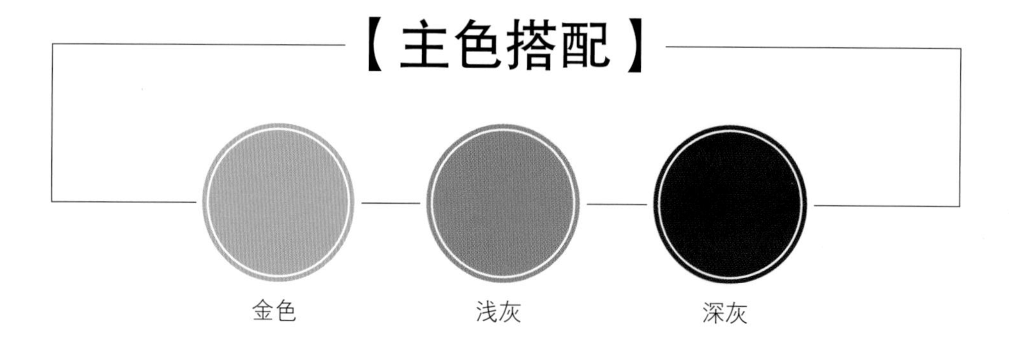

• Area / 占地面积: 1,500 m^2 • Date of Completion / 竣工时间: 2012 • Interior Design / 室内设计: HBA • Photography / 摄影: HBA • Client / 客户: IHG •

Leonardo da Vinci

FASHION NOW

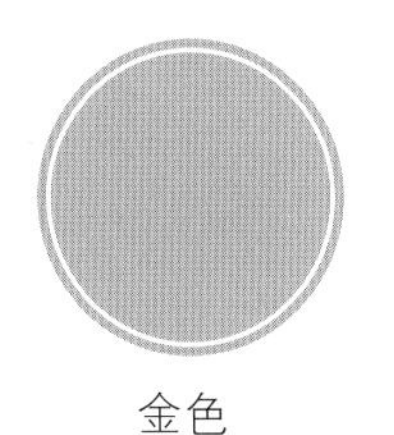

金色

浅灰

深灰

【主题色彩含义】：华丽、尊贵、辉煌

金色搭配灰色，点缀少量黑色，是打造古典主义气质最经典的配色方案之一。中庸内敛的灰色使灿烂的金色看上去高贵而不再浮躁，同时金色也使原本暗淡的灰色显得更加明亮、优雅。

【细部解析】

墙上的壁纸、软包饰面、沙发、地毯等都采用灰色系，在统一的色彩背景中形成层次丰富的质地对比。沙发靠垫的花纹、墙面上的铆钉、饰面边缘的线条和灯饰都以金色进行点缀。

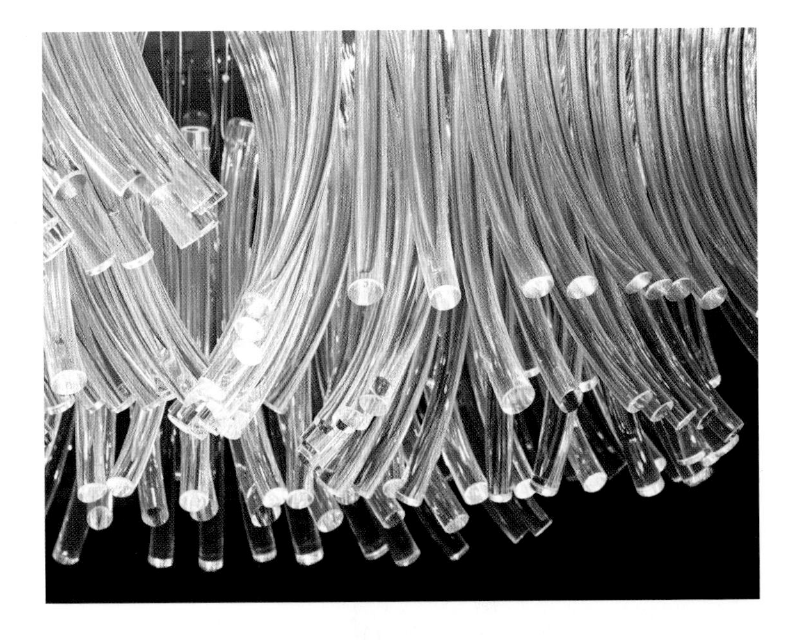

1 | 3
2

【细部解析】

1、起居室上方水晶灯细节。

2、高档玻利瓦尔灰漆饰面的质感在光线的映衬下更显尊贵。

3、深灰色背景的书房里是古香古色的书柜，上面陈列着书籍。

Grand Hotel des Iles Borromees

德斯伊利斯波若梅斯大酒店（意大利）

德斯伊利斯波若梅斯大酒店位于特斯雷萨镇上的马焦雷湖岸边，将奢华、高雅、传统与意大利高端酒店的魅力融为一身。这座由设计师成功打造的天堂酒店位于米兰附近，意大利北部的两个重要的机场—马文朋萨机场和利纳特机场为到达酒店提供便利交通。

德斯伊利斯波若梅斯大酒店在意大利奢华五星级酒店中久负盛名。酒店设有131间双人房间、25间小套房、5间单人房间、11间套房（包括海明威套房在内）。酒店所有房间为宾客提供水疗按摩及快速网络连接等服务。除此之外，宾客可以享受私人直升飞机、游泳池、网球场的便捷服务。酒店提供高标准的优质服务，具有专业的员工团队，因此也为国际会议的召开提供良好的场所。酒店著名的Centro Benessere具有专业的医疗团队，在两至七天内为宾客提供医疗服务。特斯雷萨也为体育爱好者提供航海、骑马、高尔夫等运动项目。

本案集古典传统元素与现代舒适便利于一身，堪称历史遗产，下榻德斯伊利斯波若梅斯大酒店意味着宾客可以尽享酒店的魅力、氛围及魔力。

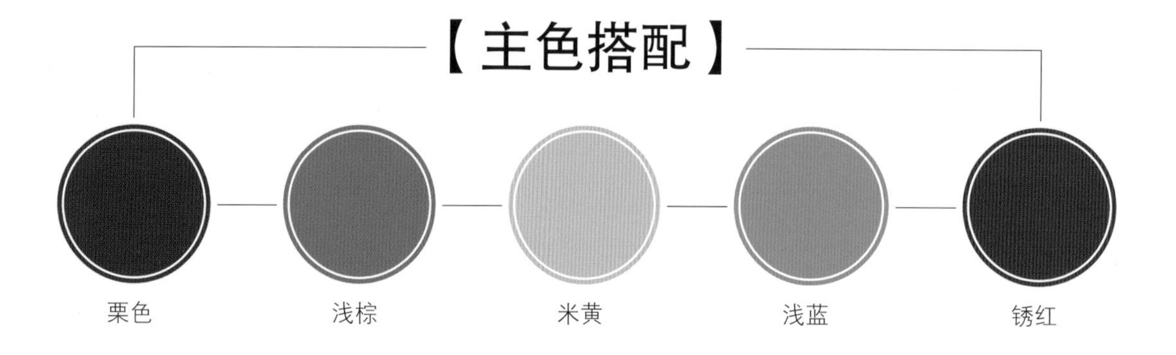

• Area / 占地面积: 33,500 m² • Date of Completion / 竣工时间: 2012 • Architecture Design / 建筑设计: Arch. Statilio Ubiali • Interior Design / 室内设计: Arch. Statilio Ubiali • Photography / 摄影: Grand Hotel des Iles Borromees • Client / 客户: S.I.A.L.M. s.r.l. Piazza Castello, 9 - Milano Italy •

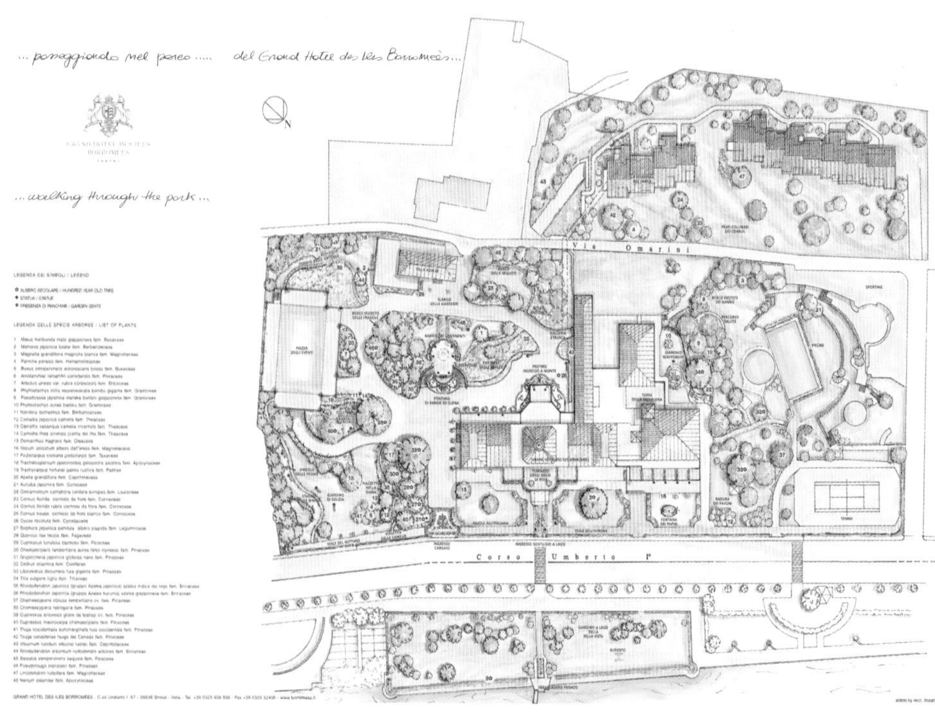
...passeggiando nel parco..... del Grand Hotel des Iles Borromées...
GRAND HOTEL DES ILES BORROMÉES
...walking through the park...
LEGENDA DEI SIMBOLI / LEGEND
Via Omarini
Corso Umberto I°

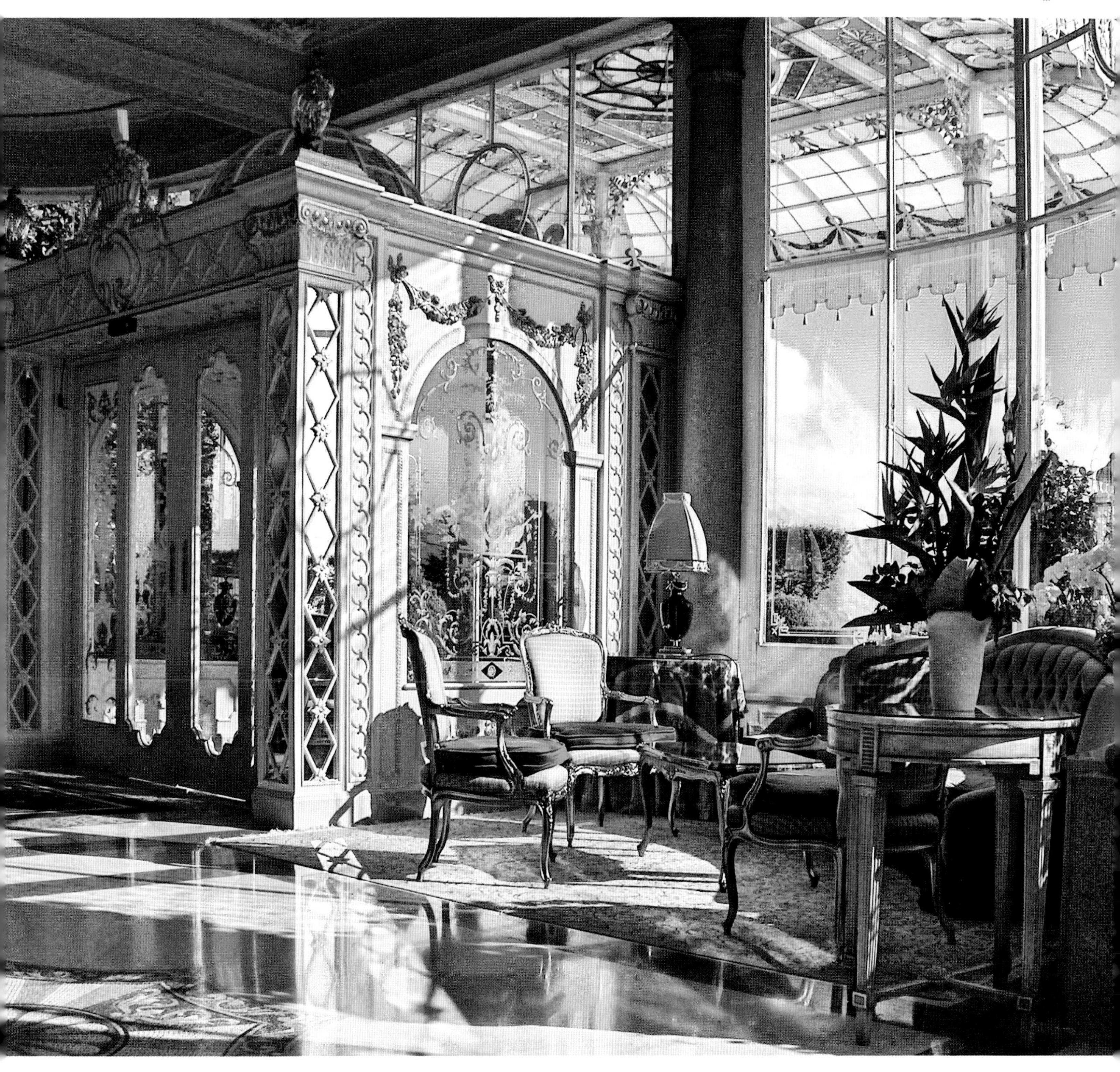

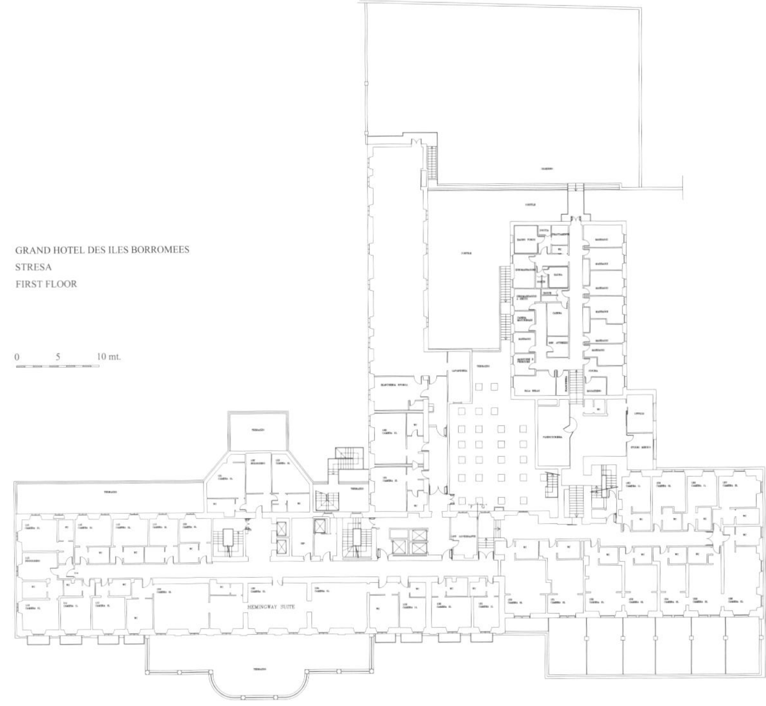
GRAND HOTEL DES ILES BORROMEES
STRESA
FIRST FLOOR
0 5 10 mt.
HEMINGWAY SUITE

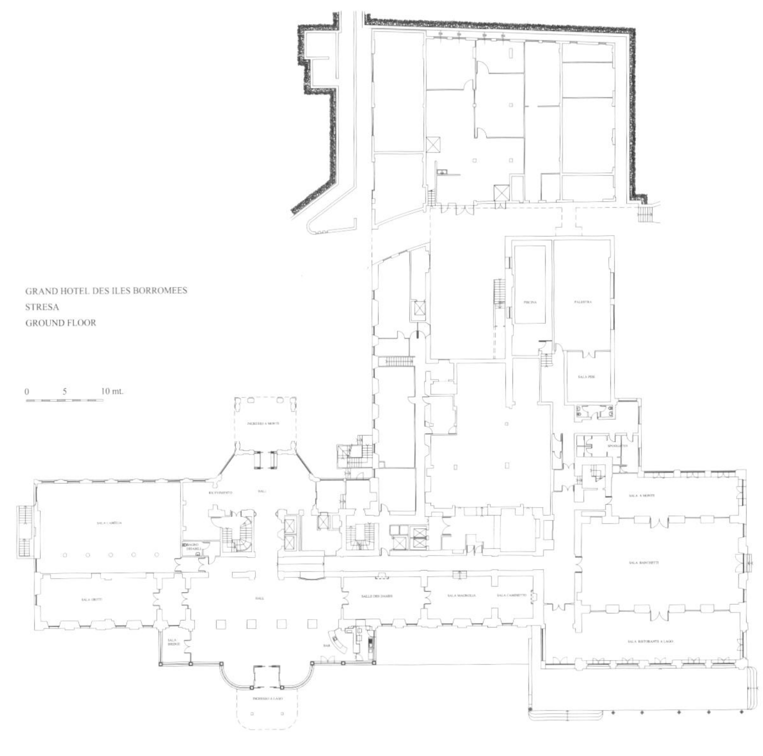
GRAND HOTEL DES ILES BORROMEES
STRESA
GROUND FLOOR
0 5 10 mt.

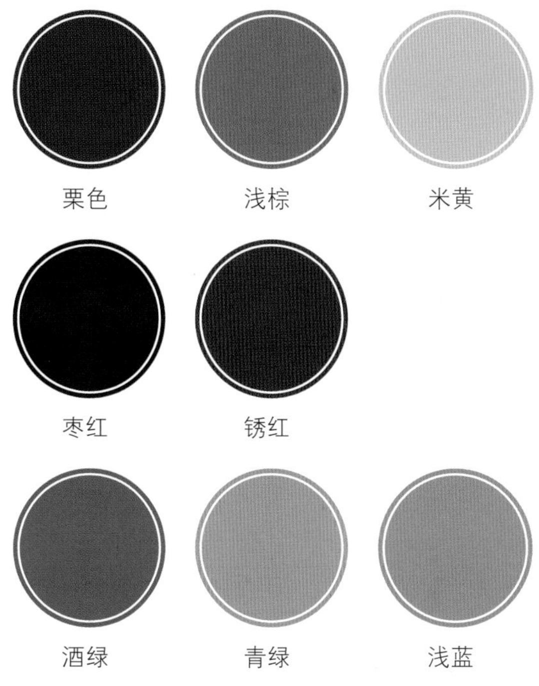

【主题色彩含义】：缤纷、丰茂、华美

分裂补色配色法（即在色轮中把一种颜色与和它垂直对应的补色左右两侧的颜色进行搭配）是一种三向配色方法，是室内配色中较复杂的一种。通过纯度和明度的变化，可以使色调更加丰富而饱满。该套房中，多种富于变化的不同纯度的红、橙和青三色系颜色互相调和的搭配，呈现出缤纷华美、精彩纷呈的感觉。

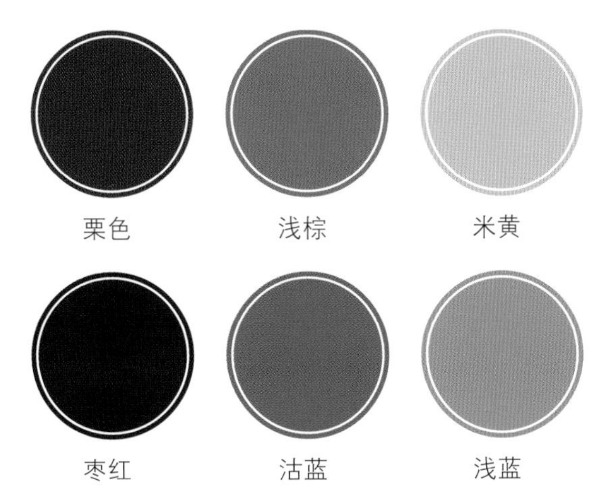

【主题色彩含义】：缤纷、丰茂、华美

同样是红、黄和青三向配色法，以棕黄色系为主色调，搭配枣红、沽蓝和浅蓝，冷色和暖色在房间中建立起既冲突又协调的对比，生动别致，又不失高贵。

GRAND HOTEL DES ILES
BORROMEES

【细部解析】

复古的台灯、巨大的枝型水晶吊灯、图案精美的壁纸和地毯，以及雕刻复杂的饰面和建筑细部，都笼罩在一片香槟色的光芒中，金碧辉煌，尽显奢华尊贵之感。

GRAND HOTEL DES ILES
BORROMEES

Taleon Imperial Hotel
塔里昂帝国酒店（俄罗斯）

坐落在圣彼得堡的塔里昂酒店，是该地区为数不多的可让食客在建于19世纪的富丽堂皇的酒店中用餐的地方。该酒店建在该宫殿用于学习和招待客人的房间，Stepan Eliseev是革命之前这座宫殿的最后一位主人，这里保留了节庆时期的壮丽景象。

位于酒店左面的前接待室被装修成了路易十六时期的风格。位于酒店右面的前书房被打造成了拿破仑波拿巴时期的帝国风格。在房间的空置处摆放着埃及的维多利亚雕塑，与此同时，壁炉上的埃及女性形象也被用于提升房间的整体美感。

塔里昂酒店是你在俄罗斯圣彼得堡地区就餐的首选地点。Alexander Dregolsky是遐迩闻名的最高级厨师。他有着卓越的能力去阐释并改良欧洲传统食物，使其充满国际化。宾客们非常满意塔里昂酒店无可挑剔的服务、种类丰富的酒水、一系列珍藏的白兰地以及最好的古巴和多米尼加雪茄。

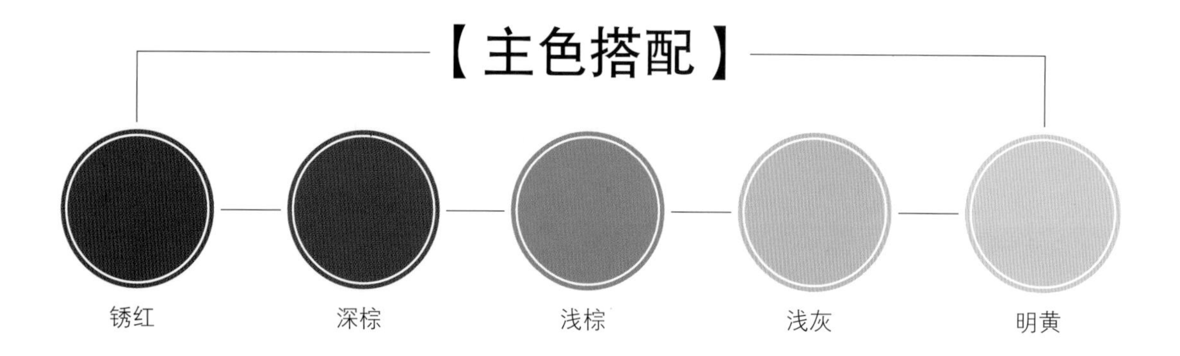

• Area / 占地面积: 27,000 m² • Architecture Design / 建筑设计: Andrew Svistunov •
Photography / 摄影: Elena Nagaenko, Vladimir Grigorenko, Philip Beloborodoff, Juriy Molodkovets, Carsten Sander,Kris Baun, Ivan Sorokin, Vladimir Stepanov •

ПЛАН 1-ГО ЭТАЖА / 1st Floor Plan

1. Главный вход
   Main Entrance
2. Стойка регистрации
   Reception Desk
3. Центральная лестница
   Main Staircase
4. Гардероб
   Cloakroom
5. Лифты
   Lifts
6. Женский туалет
   Ladies WC
7. Мужской туалет
   Gentlemen WC
8. Клубная лестница, клубный вход
   Открыто с 12:00 до 24:00
   Club Staircase, Club Entrance
   Open from noon till midnight

1. Бар «Атланты»
   Atlantes Bar
2. Гастрономический бар «Грибоедов»
   Gastronomic Bar Griboedov

1. Vertu, Tag Heuer
2. Nespresso
3. Констанс Банк
   Constance Bank
4. Сувенирный салон
   Souvenir Shop
5. Бутик элитной недвижимости "Мир квартир"
   "MK Elite" Real Estate Boutique
6. Сувенирный бутик "La Petite Opera"
   Souvenir Boutique "La Petite Opera"

Большая Морская ул.
Bol'shaya Morskaya Street

Наб. реки Мойки
Moika River Embankment

ПЛАН 2-ГО ЭТАЖА / 2nd Floor Plan

1. Центральная лестница
   Main Staircase
2. Лифты
   Lifts
3. Бизнес-центр
   Business center
4. Лифты Атриума
   Atrium Lifts
5. Винтовая лестница
   Spiral Staircase
6. Туалеты
   WC
7. Баня Елисеева
   Eliseev's Banya
8. Клубная лестница
   Club Staircase

1. Атриум-кафе
   Atrium Café
2. Ресторан «Талион»
   Taleon Restaurant

1. Салон красоты Ларисы Казьминой
   Beauty Parlour by Larisa Kazmina

201-217 Номера
Rooms

Чтобы посетить завтрак в ресторане «Виктория», поднимитесь на лифте Атриума на 6 этаж

Please, use the Atrium Lifts and go up on the 6th floor to visit Breakfast at the Victoria Restaurant

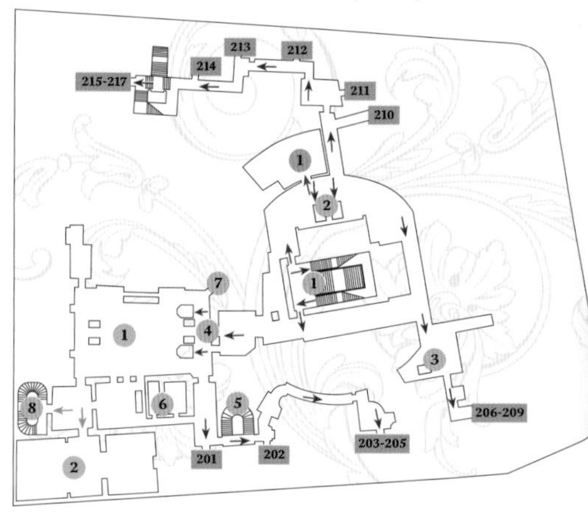

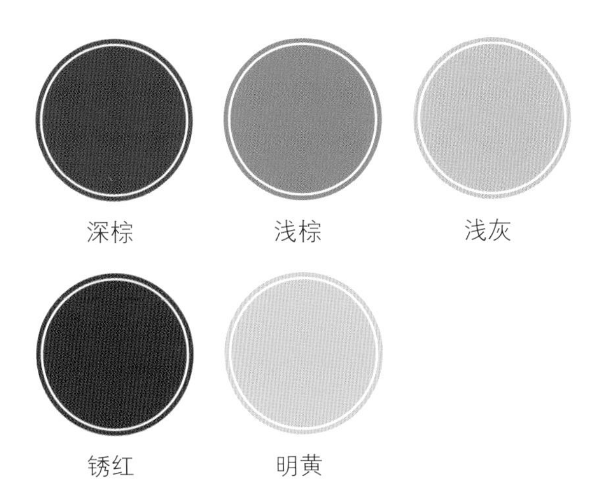

【主题色彩含义】：厚重、豪迈、辉煌

大面积的红色地毯搭配灰色墙面，以及深棕色大理石地砖和浅色墙体的搭配厚重而具有历史的跳跃感，欧派气息的造型和配色相得益彰，也使整个酒店的视觉风格令人印象深刻。

【细部解析】

两块建筑原有的穹顶装饰构成了酒店餐厅的天花板。中间的那块天花板上画着贸易守护神“水星（Mercury）”和代表幸运与成功的女神。另一块天花板上画着艺术守护神“阿波罗”和他的缪斯女神以及埃拉托。天花板上的彩绘图案、墙上挂着的古典油画和精美的陈设，都展现出酒店主人史蒂芬·伊莉斯伊夫（Stephen Eliseev）对艺术和文化的热爱。

ПЛАН 3-ГО ЭТАЖА / 3d Floor Plan

1. Центральная лестница
Main Staircase
2. Лифты / Lifts
3. Женский туалет / Ladies WC
4. Мужской туалет / Gentlemen WC
5. Лифты Атриума
Atrium Lifts
6. Клубная лестница
Club Staircase
7. Туалеты / WC

1. Зал «Баккара»
Baccarat Ballroom
2. Золотая гостиная
Golden Salon
3. Музыкальная гостиная
Musical Salon
4. Библиотека
Library
5. Ореховая гостиная
Walnut Salon
6. Зал «Империал»
Imperial Grand Hall
7. Анфилада
Enfilade
8. Бар «Грот»
Grot Bar

301-313 Номера
Rooms

Чтобы посетить завтрак в ресторане «Виктория», поднимитесь на лифте Атриума на 6 этаж

Please, use the Atrium Lifts and go up on the 6th floor to visit Breakfast at the Victoria Restaurant

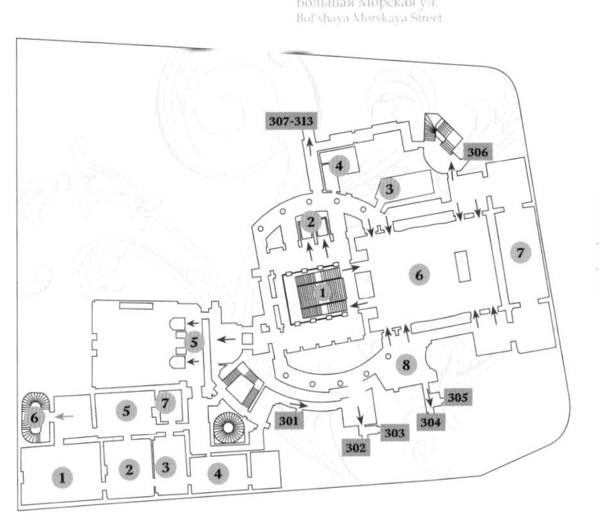

Fairmont Peace Hotel
费尔蒙和平饭店（中国）

作为中国地标性建筑之一，和平饭店的建筑风格属于芝加哥学派哥特式建筑。酒店早在1929年8月1日即开始营业，翻新后的饭店由费尔蒙酒店集团负责管理，作为“远东第一楼”闻名中外。它地处上海外滩，建筑外观和内部装饰都极其奢华精致。HBA与一队顶级设计师、建筑师和历史学家组成团队复活了这个地标性建筑的宏伟壮丽外貌，并将其注入优雅的风韵。

全新的费尔蒙和平酒店内有270个高级客房和套房，带有6个餐厅和休息室。其中包括深受顾客喜爱的爵士酒吧、20世纪30年代风格的上海风情厅、龙凤厅等。HBA对费尔蒙和平饭店的设计将唤起上海装饰艺术的回归，加之流线型家居设计以及最新式室内设施，都凸显了HBA团队的设计特色。

著名的“九国主题套房”是饭店的一大特色，其中印度、英国、中国和美国这四个国家的套房保持原样；而法国、意大利、西班牙、日本和德国套房在保留原来设计理念的基础上做了稍微的改造。和平饭店的10楼设为总统套房，这里曾经是和平饭店原创始人Victor Sassoon居住过的地方。

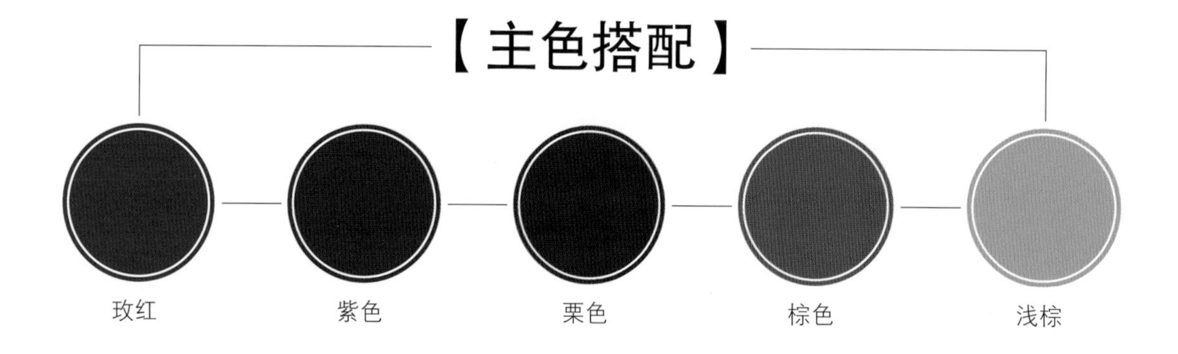

• Area / 占地面积: 1,500 m² • Date of Completion / 竣工时间: 2010 • Interior Design / 室内设计: HBA • Photography / 摄影: HBA • Client / 客户: FAIRMONT •

1	3
2	

【细部解析】

1、辉煌的八角形玻璃天窗。

2、米色意大利大理石铺成的大堂地面上饰有精美的纹理图案。

3、棚顶上悬挂的古铜镂花吊灯豪华典雅。

【细部解析】

复杂的建筑角线、高品质的装饰面板和奢华的灯饰，这些精致的细节无一不彰显着古典奢华酒店设计及施工的高标准要求。

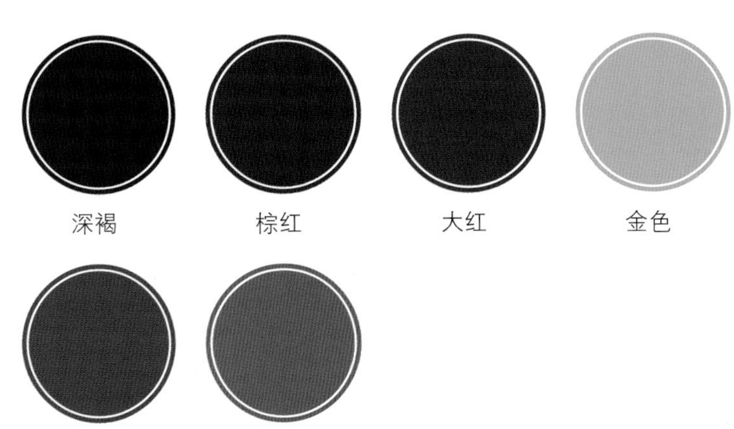

【主题色彩含义】：灿烂、复古、奢华

建筑背景和餐桌、餐椅都采用金黄色，周围墙面上方以紫色灯光进行渲染，形成了强烈的补色对比效果。再搭配棕红色的地板，同时加以少量灰色和红色的布艺点缀，使餐厅的色彩趋于华丽而高贵的感觉。

【细部解析】

沙发、椅子等家具选用的材料都十分考究，高品质的布料、实木与真皮形成了丰富而层次分明的材质对比。图案精美的地毯和大理石地面搭配得相得益彰。

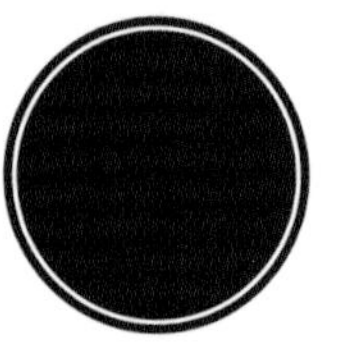

深褐　红棕　朱红　金黄　灰绿　米黄

【主题色彩含义】：丰富、复古、典雅

古香古色的中式古典家具里面陈列着价值不菲的玻璃艺术品，地面上铺设的地毯上是色彩斑斓的几何色块图案。现代与古典的元素在富有节奏的色彩搭配中协调为一体。

【细部解析】

上图是意大利主题套房。套房中的会客厅、餐厅、卧室和浴室的设计都十分考究。窗帘、靠垫和地毯上的古典精美的装饰图案与做工精良、造型典雅的家具，无不唤起现代意大利设计特有的精工细作的品质感。

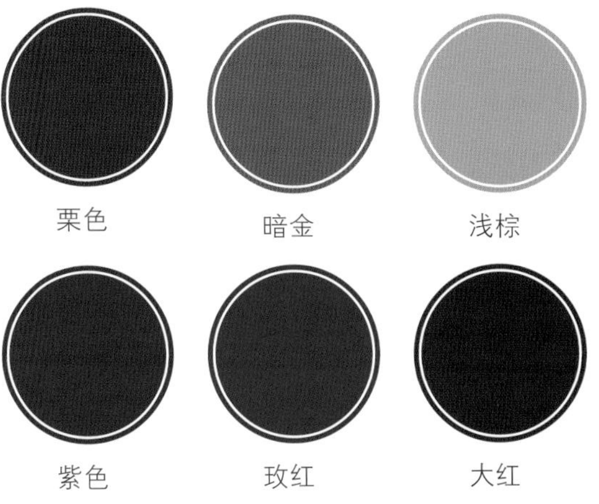

【主题色彩含义】：厚重、端庄、奢华

该套房中的配色采用了类比色设计（棕色系）与二次色设计（低明度、低饱和度的橙搭配紫色）相结合的配色方法。这种配色法比较复杂，呈现的视觉效果也给人层次丰富、厚重绵长的感觉。

【细部解析】

宽大、舒适的实木家具在房间内对称摆放。家具的雕刻细节上摒弃了过多的繁琐与奢华,呈现出经典的美式优雅风格特色。随处可见的Art Deco装饰派艺术更是令人印象深刻。

【细部解析】

英国套房采用橡木镶板作为墙面饰面板，上面的雕刻花纹十分细腻精美。桌面上的一大束玫瑰花和橡木树叶的天花板雕刻，都突出了维多利亚时代的英格兰风格印迹。

【细部解析】

印度套房的设计强调了富贵华丽的特点。印度特色的花纹装饰配以明丽的颜色打造出清真寺似的穹顶，下面挂着奢华的吊灯。地板上采用了精美的掐丝石膏工艺设计，地毯和家具、靠垫的图案都非常具有印度特色。

【细部解析】

法国套房内采用了艺术运动风格的古典墙纸进行装饰。木质地板配深色厚重的地毯，以及胡桃木雕刻家具和豪华考究的布艺饰品，都呈现出浓厚的古典浪漫气息。

Four Seasons Nile Plaza at Cairo

开罗尼罗河四季酒店（埃及）

该案是开罗尼罗河广场四季酒店内著名的私人住宅的室内设计。该案位于酒店私人套房区28楼，公寓占用了整层，面积近1000m^2。本案所在的商住综合楼俯瞰开罗市内的尼罗河，临近繁华的马路。河岸附近有着多样的阳台和咖啡厅，在高处可遥看金字塔。鉴于业主繁忙的商务活动需要一个有名气的地点用于会见商业伙伴，大部分的公寓空间主要用于商业活动，如总统办公室，配有电话会议设备的会议室，适合机要会议的休息室和员工休息空间。私人生活区保留了雅致奢华的风格，这也是整体室内设计的主题。整体室内设计的主要元素是木制镶板，所有的墙面都用这种精致雕刻的金色基调的镶板包裹，这也与天花板的设计相呼应。

地面的材料主要为意大利锡耶纳黄色大理石，穿插彩色镶边和次等宝石。部分地面与房间相连，轻质木地板和精雕的地板全部产自意大利。室内家具全部出自意大利当代设计大师之手，与室内的古典风格完美地融为一体。沙发和休息室内的扶手椅及室内家具的皮面和绒面都是意大利著名品牌。布艺饰品的颜色与墙壁饰面板的象牙色呈现鲜明的对比。其他的重要设计细节如吊灯、慕拉诺（Murano）玻璃、带有叶片和装饰的水晶以及丝质灯罩，带给室内鲜明的设计符号。业主希望被公寓内的珍贵器物环绕，因此要求配件、挂画、雕塑和地毯等陈设饰品全部为独特设计。

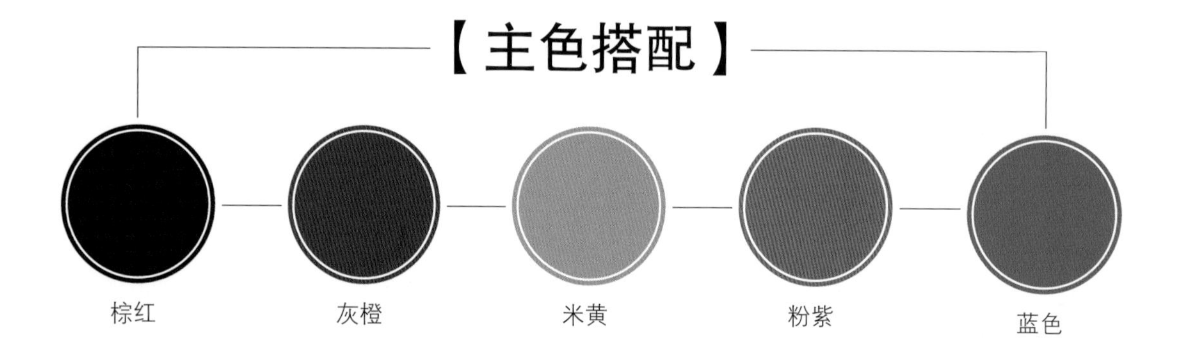

• Area / 占地面积: 1,000 m^2 • Date of Completion / 竣工时间: 2011 • Architecture Design / 建筑设计: Cleopatra Group Arch. Dpt •
Interior Design / 室内设计: ARCHITETTURAMBIENTE • Photography / 摄影: Arch. Maddalena Bosi •

1	3
2	

【细部解析】

1、精美的蓝色玻璃饰品起到画龙点睛的作用。

2、雕刻精致烦琐的家具呈现出浓郁的古典主义风情。

3、地面的材料主要为意大利锡耶纳黄色大理石，中间嵌有彩色镶边和次等宝石。

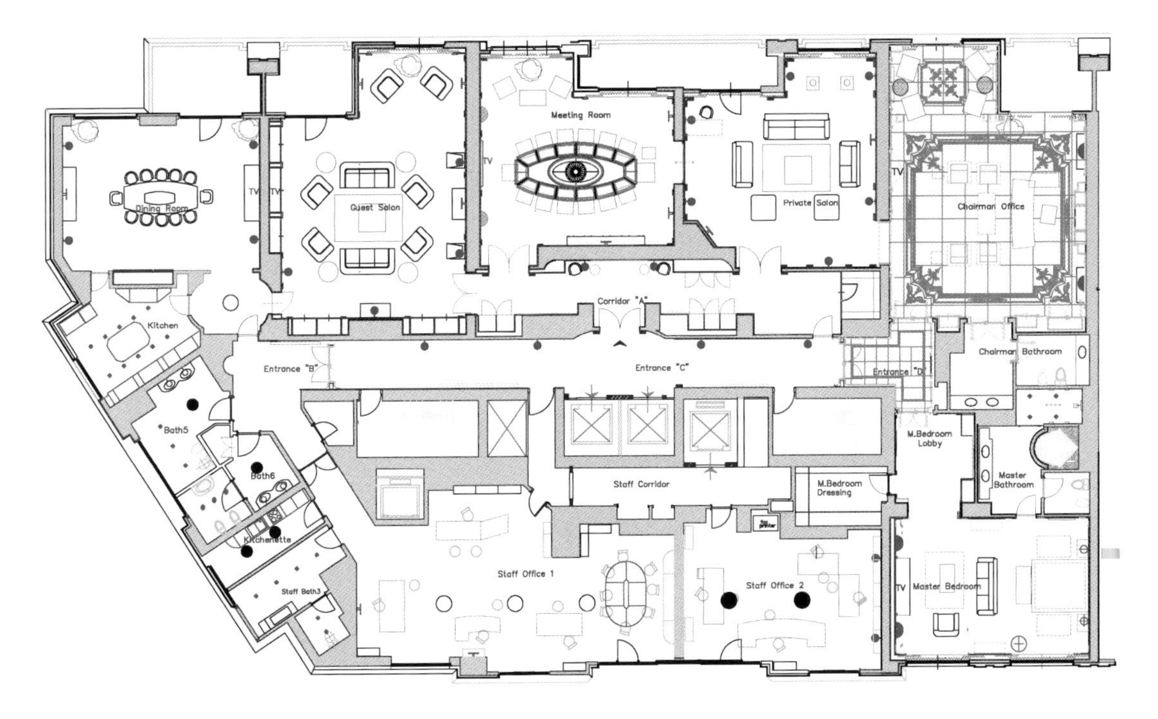
Meeting Room
Dining Room
Guest Salon
Private Salon
Chairman Office
Corridor "A"
Kitchen
Entrance "B"
Entrance "C"
Entrance "D"
Chairman Bathroom
Bath5
Bath6
M.Bedroom Lobby
Master Bathroom
Staff Corridor
M.Bedroom Dressing
Kitchenette
Staff Office 1
Staff Office 2
Staff Bath3
Master Bedroom

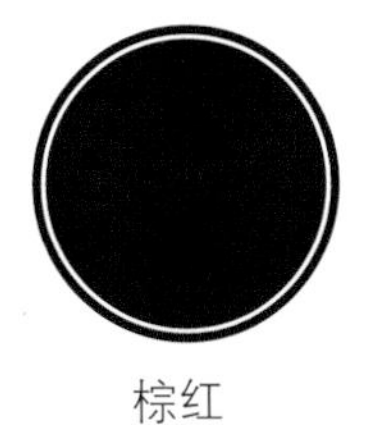
棕红

灰橙

米黄

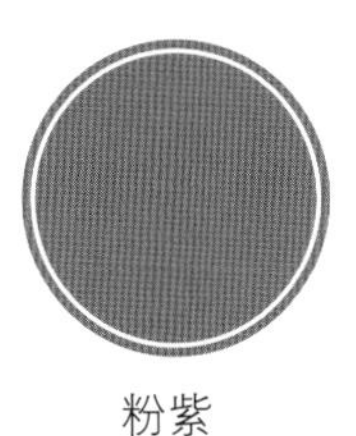
粉紫

【主题色彩含义】：柔美、温暖、光辉

米黄色调的墙面背景、沙发、脚踏、皮草垫和窗帘布艺色彩统一而优雅，同时搭配橙红色调的地板和家具使房间显得温暖舒适。紫色的灯饰更为客房增添了柔和典雅的气氛。

Gaylord Texan Resort & Convention Center

盖洛德德克萨斯度假酒店及会议中心（美国）

设计要求通过室内和建筑设计体现德克萨斯州各地区和文化的丰富多样。设计团队谨慎的处理“典型的德州”这一主题，希望为这个世界顶级会议酒店打造出优雅好客、充满该州美丽自然风光的空间。

德州地域广袤的特点体现在建筑材料全部来自本州而且丰富多样。大理石、花岗岩和石灰石是室内空间的主要元素。为平衡墙面和地面，设计团队采用柔软的家具陈设使空间看起来更加柔和，包括有毯子包裹的椅子、大面积的地毯、质朴的铁艺和木制品，以及西方古典艺术品。质朴的土红色、巧克力棕、浅棕和大胆的松石绿色点缀，都为酒店增添了西式精致感。

酒店大堂体现了真实的“德州”风光。德州墙石、橡木天花板和孤星长角图案营造出华丽的入口空间，这种感觉一直延续到接待前台，客人可以在此办理入住手续。前台的设计硬朗且雕刻感十足，主要选用了大理石和实木两种材料。大堂的长廊里有一个巨大的石壁炉，欢迎客人加入到银星酒吧放松心情。奢华细致的手工地毯，以红色、棕色和蓝色的鲜明对比增添舒适、放松的柔和感。大厅中精雕细刻的西式家具，凸显高端的“德州风格”。皮质家具垫缝纫细致，手工上色，采用绿松石色和中性色调的铆钉固定和装饰。

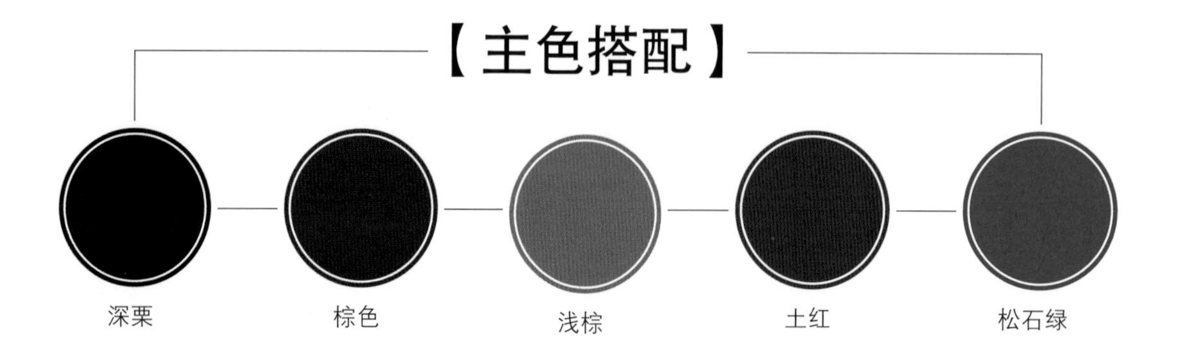

• Area / 占地面积: 21,3677 m^2 • Date of Completion / 竣工时间: 2004 • Architecture Design / 建筑设计: Hnedak Bobo Group •
Interior Design / 室内设计: Wilson Associates • Photography / 摄影: Michael Wilson, Dallas, TX •

1	3
2	

【细部解析】

1、前台的设计硬朗且雕刻感十足，主要选用了大理石和实木两种材料。

2、家具陈设点缀在石材空间中，使整体氛围看起来更加柔和。

3、大理石、花岗岩和石灰石是室内空间的主要材料和元素。

【细部解析】

“Amalur”餐厅吧台的马赛克拼花图案设计十分精美而独特。吧台上红色烛台、酒杯和吧椅的红色皮面形成呼应，同时与松石绿和黄色形成对比，使餐厅吧台产生了强烈的视觉冲击力。

GRAVES and MEINZER
TEXAS HILL COUNTRY
COWGIRLS

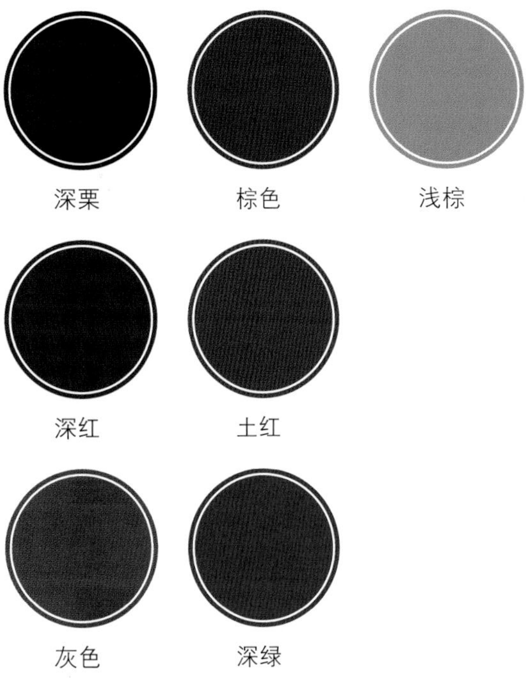

【主题色彩含义】：稳重、信赖、感染力

棕色是土地的颜色，体现着广泛存在于自然界中的真实与和谐，所以是让人觉得平和稳定并带给人安全感的颜色。棕色代表稳重和中立，也充满生命力。大面积使用棕色系，同时以深红和土红提亮活跃氛围，并点缀绿色植物的配色方法非常具有感染力。

Hotel Alfonso XIII
阿方索十三世酒店（西班牙）

西班牙南部古都塞维利亚珍贵的地标建筑阿方索十三世酒店为了巩固其作为欧洲顶级豪华酒店的显赫地位，进行了重新装修。HBA伦敦工作室的The Gallery设计师独具匠心，揉合真实历史史料及以塞维利亚为中心的安达卢西亚文化，同时秉承酒店的独特魅力，将故事娓娓道来。

酒店最初是由西班牙国王阿方索十三世下令建造，于装饰艺术鼎盛时期的1929年开业，在当时的“旅游黄金时代”吸引了众多旅客驻足停留。塞维利亚是安达卢西亚文化的中心城市，当地曾被摩尔人统治500年，是斗牛与佛朗明哥舞蹈的起源地，也是西班牙家喻户晓的兼男性阳刚与女性神秘魅力于一身的情圣唐璜的故乡。HBA从这些独特元素中汲取灵感，细细道出历史故事，保留并发扬了古韵，同时增添了现代的新意，打造出与时俱进的“豪华精选”酒店。

三种迥然风格的客房设计分别融入塞维利亚最重要的摩尔（Moorish）、安达卢西亚（Andalucian）以及卡斯蒂利亚（Castilian）三种文化：“摩尔式客房”采用复杂精细的古典装饰线条，摆放时尚新潮的家具及各种造型优美的摆设；“安达卢西亚客房”从佛朗明哥舞蹈中汲取灵感，天花线雕刻的柔美曲线令人不禁浮想起舞裙的摇曳风姿，明艳而具有动感，并搭配细碎花纹的纺织面料的华丽皮革床头板，整体装饰女性魅力十足；“卡斯蒂利亚客房”则散发如同斗牛士在竞技场上挥舞斗篷奋战时的阳刚之气，客房采用深赭石色为主调，在其他鲜亮色彩和深色木质家具，如精心雕刻的床头板的映衬下显得更为迷人。房间缀以用笔大胆奔放的画布，更营造出强烈的戏剧感。

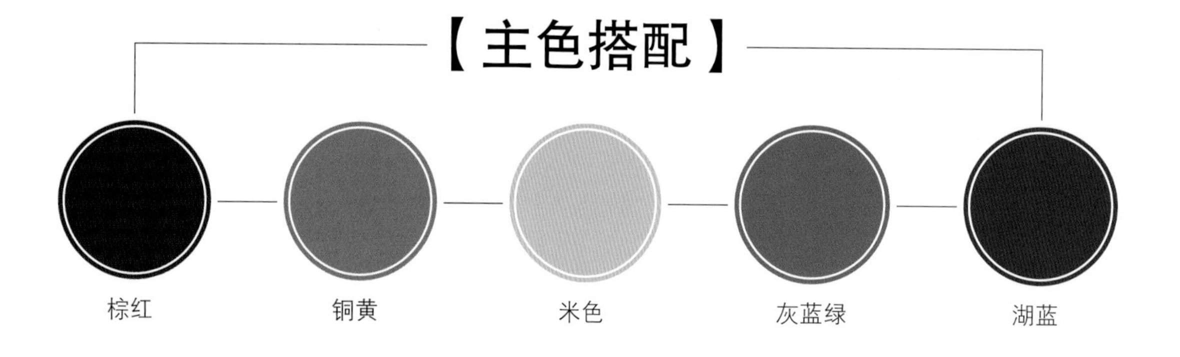

• Area / 占地面积: 1,500 m² • Date of Completion / 竣工时间: 2012 • Interior Design / 室内设计: HBA • Photography / 摄影: HBA •

【细部解析】

酒店的创新设计无不体现出塞维利亚的当地特有风格。优雅的线条与传统的建材巧妙融合，营造出舒适质朴而又不失优雅的迷人氛围。

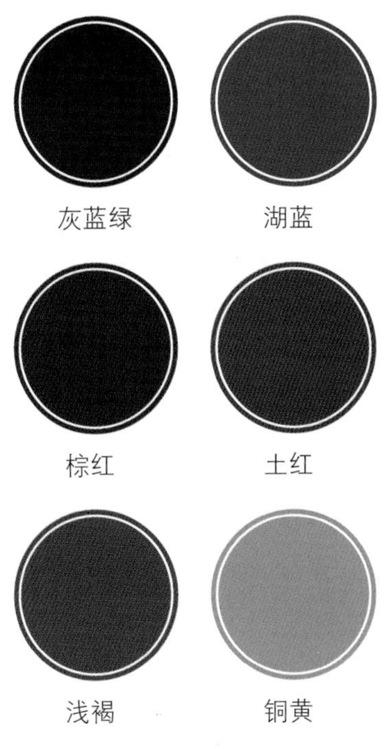

【主题色彩含义】：装饰艺术、摩登

光漆墙身搭配湖蓝色的巨大镜框和灰蓝绿色丝质织布窗帘，与棕红色的窗框、土红色的座椅、檀木制成的浅褐色吧台以及金色闪亮的灯饰形成鲜明对比。丰富的对比色搭配使“美式酒吧”的设计重新演绎了装饰艺术风格。

【细部解析】

1、复杂精美的壁画在红黑两色构成的背景中显得古老而神秘。

2、壁画、拱门浮雕和家具陈设都散发着浓郁的塞维利亚韵味。

3、背光摩尔式雕刻屏风营造出颇为私密的用餐氛围。

1	2	3

【细部解析】

1、摩尔式客房内复杂精细的古典装饰线条搭配造型优美的陈设品。

2、安达卢西亚客房采用深赭石色为主调，端庄古雅。

3、卡斯蒂利亚客房卧室内精心雕刻的床头和深色木质家具低调奢华。

1	2	3

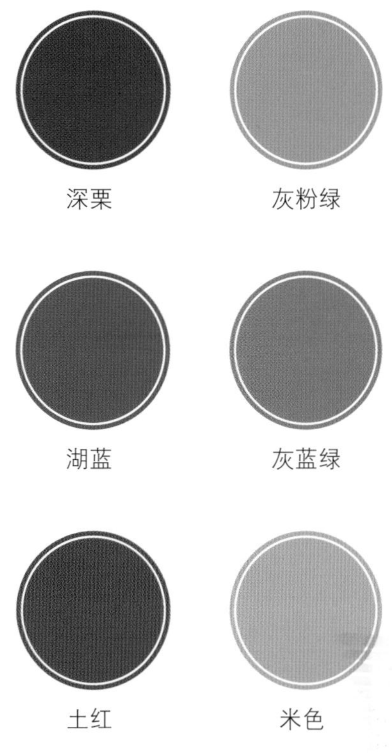

【主题色彩含义】：明快、动感

“安达卢西亚客房”从佛朗明哥舞蹈中汲取灵感，白色天花线雕刻的柔美曲线令人不禁浮想起舞裙的摇曳风姿。米色、土红、深栗色构成的背景中点缀蓝绿色的靠垫与湖蓝色的挂画，形成柔和的冷暖对比，给人明快而活跃的感觉。

1	3
2	

【细部解析】

1、皇家套房主卧室内的四柱大床上挂有精美的手工刺绣帷幔。

2、皇家套房内珍贵古董与奢华时尚设施相映成趣。

3、皇家套房的卫生间既古朴典雅又不失时尚。

【细部解析】

Reales Alcázares套房内，木炭色墙壁与沙发靠垫和地毯上的传统图案相映成趣。主卧室以厚重的深色调天鹅绒窗帘搭配精致的铁艺家具，颇有一丝神秘魅惑的味道。

Hotel Palazzo Versace
范思哲宫殿酒店（澳大利亚）

闻名于世的奢华酒店范思哲宫殿酒店坐落于澳大利亚著名旅游胜地昆士兰黄金海岸。酒店内的家具、饰品都选自范思哲家居品牌（House of Versace），明快生动的色彩、图案和时尚的风格都反映了范思哲的设计哲学，其中包括每个房间内都运用了范思哲品牌的经典符号——“美杜莎头像”。酒店的每处设施都选自范思哲品牌，这也符合该酒店设计师专注于设计五星时尚酒店的传统。

范思哲酒店可提供200套古典雅致的客房及套房，72间临近的公寓和一个直径为65米的湖泊。该酒店还包括三个曾经获奖的餐厅：“Vanitas”、“Vie”和 “Il Barocco”。它还拥有一个豪华的大堂酒吧名为“梦中的花园”，可提供库蒂尔鸡尾酒和高级的下午茶。范思哲酒店还可以提供一系列五星级标注的特色服务，包括水沙龙、健身+健康中心和极光矿泉疗养等，这些均于2012年底正式投入使用。酒店可提供令人惊叹的活动空间，包括美杜莎舞厅、圆形会议室以及喷泉平台，可同时容纳12到500人不等的团体。

伴随着壮观的水景设置，该酒店的建筑、环境布局、家具摆放以及其中的装饰物都反映出仅限于欧洲地区大型酒店的奢华感。富丽堂皇的装饰空间也显示出它已经超越了用意大利马赛克装饰，带有拱形黄金天花板和名贵古董吊灯的米兰州立图书馆的奢华程度。范思哲酒店多次被誉为澳大利亚最佳酒店，同时被评选为2013年旅游业最佳酒店/度假地，并且在2013年康德纳斯特旅行者金牌榜上荣登榜首。

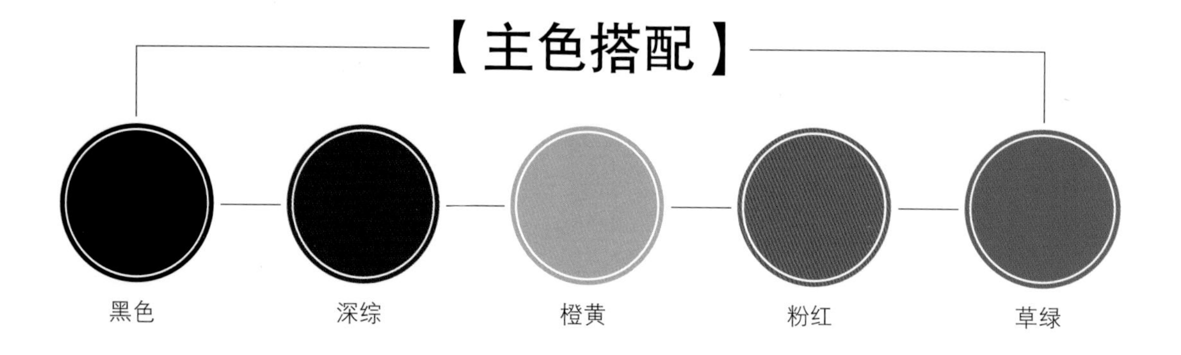

• Date of Completion / 竣工时间: 2012 • Architecture Design / 建筑设计: Rocco Magnoli • Client / 客户: Palazzo Versace Gold Coast •

vie

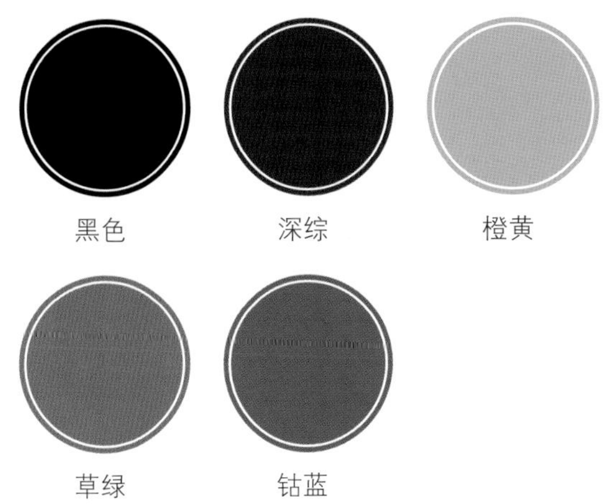

【主题色彩含义】：富丽堂皇、奢华、灿烂

棕色系与黑色相间的马赛克拼花地面与同样色调的格子形拼花大理石地面相结合，与米色背景融为一体。巨大的水晶吊灯与对比鲜明的黄蓝花纹座椅使大堂空间笼罩在富丽堂皇、灿烂夺目的光辉之中。

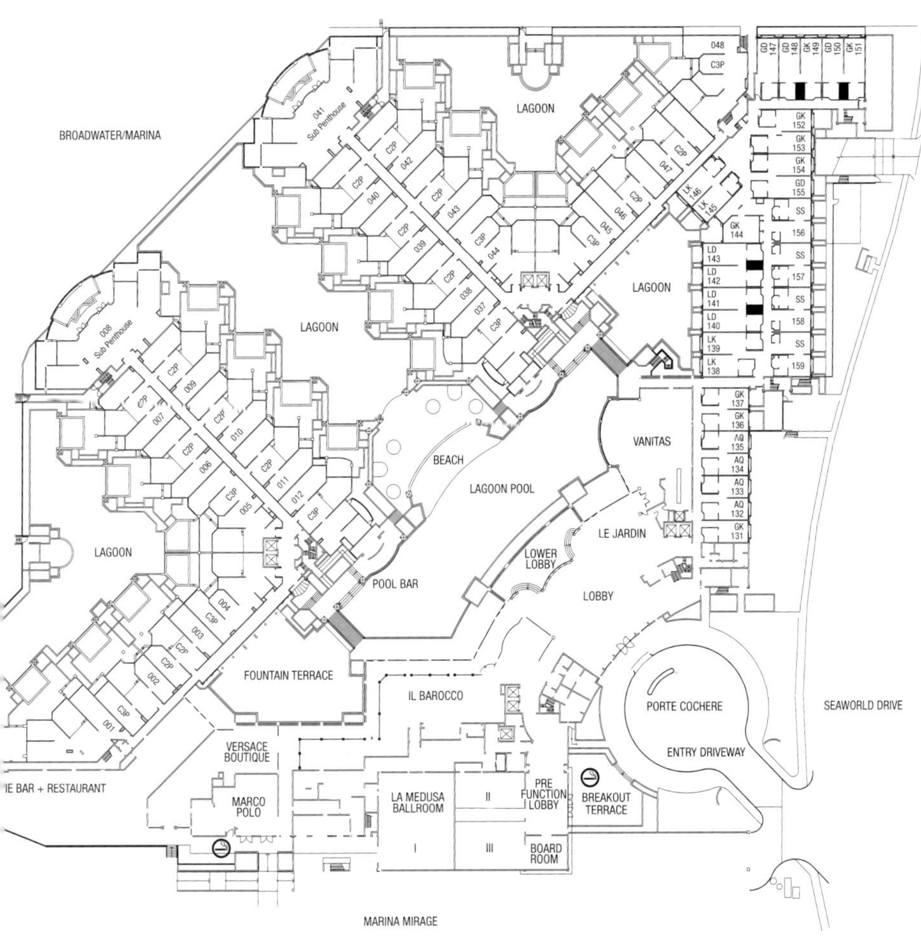

【细部解析】

酒店内的家具、饰品都选自范思哲家居品牌（House of Versace），明快生动的色彩、图案和时尚的风格都反映了范思哲的设计哲学，其中包括每个房间内都运用了范思哲品牌的经典符号——“美杜莎头像”。

【细部解析】

酒店内设有三个曾经获奖的餐厅："Vanitas餐厅"、"Vie餐厅"和 "Il Barocco餐厅"。色彩方面主要选用棕色、褐色等实木原色，搭配少量的湖蓝色作为对比。

One&Only Le Saint Géran

乐·圣热兰One＆Only度假村（毛里求斯）

早在10世纪，阿拉伯船队发现毛里求斯时便称它为“彩虹、瀑布和流星之地”。1510年期间葡萄牙人曾访问过这里。1598年荷兰人第一次占领了此地，并以他们国家领袖穆里斯王子的名字命名。

度假村的166个套房和一栋别墅都与最新科技完美融合在一起，目的是提升住宿的品质。所有套房均设有露台或者阳台，面朝海洋或海湾，并且通向热带风情的花园。洁白的床单全部为埃及纯棉布，配以同色的鹅绒软枕和灰色格子图案的床盖，以及米黄色系的家具，巧妙的色彩搭配充满了印度洋风情，也为房间增添了一抹温和的氛围。宽敞的浴室布置美观，配备大型的层叠淋浴喷头。所有客人都可享受个性化的24小时管家服务。

作为国际级水疗中心，One&Only与ESPA亲密合作打造出放松和舒缓的极致体验。私人小型泳池的安静避风湾，该水疗中心提供全套服务包括全身护理和个性化的健身项目以及营养分析。水疗中心也提供全套美容服务以及Bastien Gonzalez足疗护理。

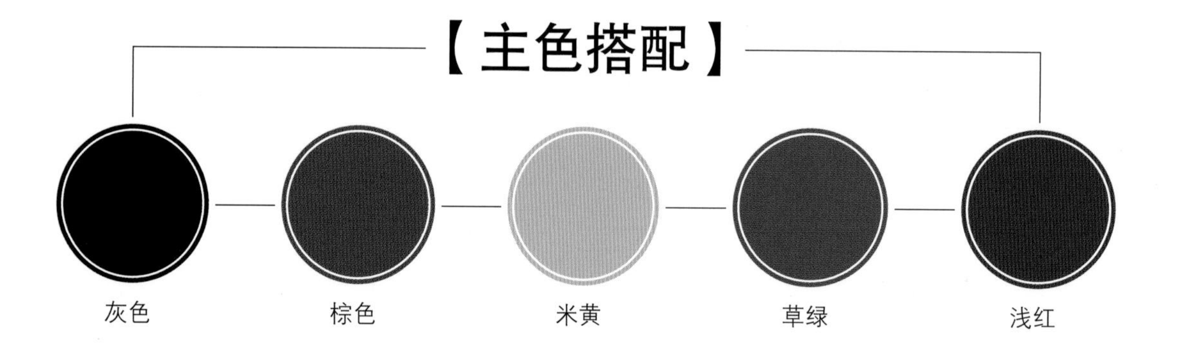

• Date of Completion / 竣工时间: 2013 • Design Firm / 设计团队: One&Only Le Saint Géran • Photography / 摄影: One&Only Le Saint Géran • Client / 客户: One&Only •

【细部解析】

硬朗的石材和温润的实木作为主要装修材料，与布艺沙发优雅柔软的面料形成鲜明对比，同时搭配暖色调的古典图案地毯以及精致的花束、花篮、灯具等饰品，让就餐区充满了闲适、惬意的度假气息。

灰色

棕色

米黄

草绿

浅红

【主题色彩含义】：愉悦、温和、印度洋风情

所有套房均设有露台或者阳台，面朝海洋或海湾，并且通向热带风情的花园。洁白的床单全部为埃及纯棉布，配以同色的鹅绒软枕和灰色格子图案的床盖，以及米黄色系的家具，巧妙的色彩搭配充满了印度洋风情，也为房间增添了温和的氛围。

One&Only Palmilla

One＆Only帕尔米亚豪华度假村酒店（墨西哥）

One&Only帕尔米亚豪华度假村酒店位于加利福尼亚半岛最南部角落的天堂之地，乃墨西哥洛斯卡布斯区(Los Cabos)豪华度假村的先驱。

这家海滨酒店拥有带spa理疗的13间私人别墅和一个27洞高尔夫球场，并提供享有太平洋或Cortez美景的客房。One&Only帕尔米亚豪华度假村酒店的客房和套房享有墨西哥风格设计。在这里，古老墨西哥的优雅风格，以红瓦屋顶和白灰墙的形式体现出来。闲适的喷泉和摇曳的棕榈树相映成趣，使这里俨然成为了鸟类的天堂，热带花儿竞相绽放。

One&Only帕尔米亚豪华度假村酒店秉承了One&Only品牌致力在全球发展最好酒店的传统，是美洲沿岸的又一旗舰店，为当地追求品质与私密体验的客人提供了一个全新的选择。酒店每间客房和套房均设有庭院或阳台，并配备了平面电视和iPod基座。这家酒店被列为世界前100酒店及度假村。

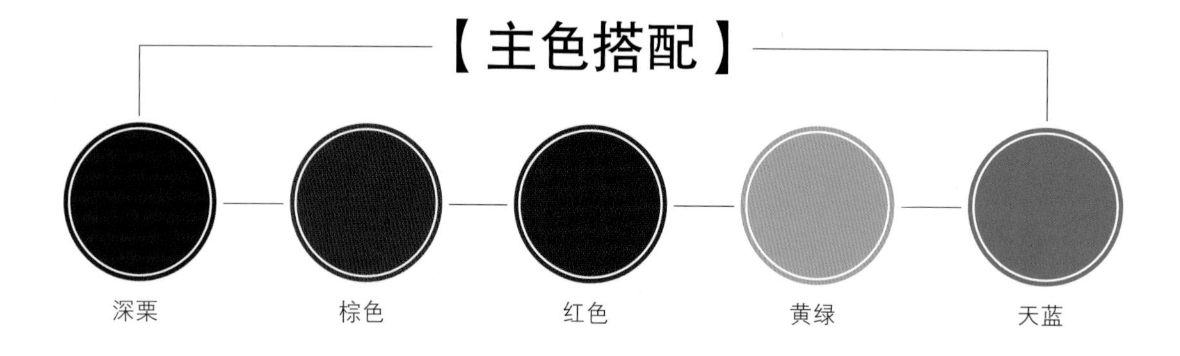

• Date of Completion / 竣工时间: 2013 • Design Firm / 设计团队: One&Only Palmilla • Photography / 摄影: One&Only Palmilla • Client / 客户: One&Only

【细部解析】

古老墨西哥的优雅风格，以红瓦屋顶和白灰墙的形式体现出来。闲适的喷泉和摇曳的棕榈树相映成趣，鸟鸣花妍相得益彰，使这里成为人们衷爱的度假天堂。

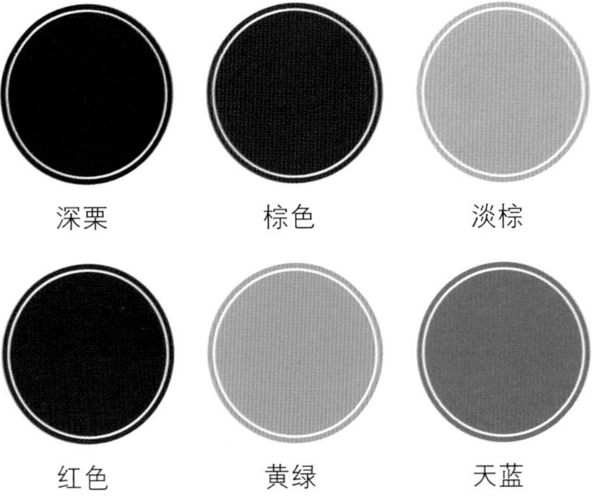

【主题色彩含义】：愉悦、闲适、晴朗

深栗色的实木床、棕色的天棚角线、家具、镜框、大理石地板等构成了棕色系的大基调，小面积点缀鲜艳明亮的红色装饰、黄绿色躺椅，在窗外蓝色海面的映衬下，显得清爽洁净，带给人舒适愉悦的感受。

The Alpina Gstaad

阿尔皮纳·格施塔德酒店（瑞士）

阿尔皮纳·格施塔德酒店集传奇色彩、优越地理位置及瑞士传统于一身，兼具时尚和丰富多样的别致特色，被誉为“新经典的诞生”。这原本是一幢极为特别的私人宅第，洋溢独具匠心的奢华与个性；设计师通过汲取其悠久历史重新进行构思，并注入对豪华酒店未来发展朝向的深入洞悉，令其变身为一座与众不同的奢华酒店。

壁炉是酒店致力为宾客创造当地特色住宿体验的重要体现。壁炉周围的墙面由阿尔卑斯岩石建成，每一块都是从当地的河流里精选而来，经过数世纪激流的冲刷打磨。另外一个大小相若、同样由岩石堆砌而成的壁炉位于楼上的酒廊，壁炉连接了两个空间，实现了怡人的流畅感。两个楼层之间的宽敞楼梯既具有当代风格，细节中亦体现出完美精致的瑞士传统。楼梯为木质，扶手和栏杆杆则分别由白煤钢和玻璃打造，之后由工匠于现场在扶手表面缝上最高级的鞍皮。细致描画、流露古雅色彩的天花板为宾客呈现一场视觉盛宴，是该酒店绝无仅有的一处特色。在楼梯后方，板条木屏风借鉴当地Saanenland农场的“gimmwand”建筑风格，形成优雅的栅格，使投射下来的灯光影影绰绰，份外柔和。

HBA的高级联系总监Nathan Hutchins表示：“突出个性是我们设计每一个项目的理念。我们从各个项目所在地的历史、文化及自然环境入手寻找其个性，然后赋予它们与环境完全契合的全新生命力。The Alpina Gstaad就是一个完美的例证——这家新酒店既传承了当地的极致工艺，亦具有与众不同的低调奢华。”

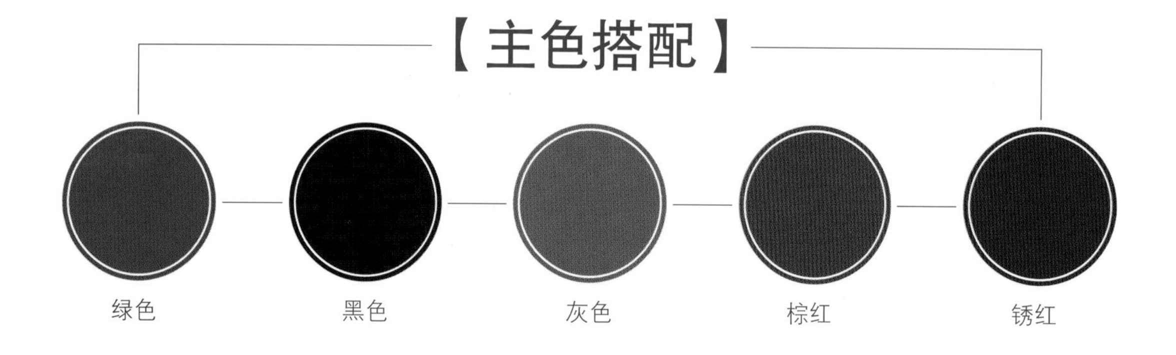

• Date of Completion / 竣工时间: 2012 • Interior Design / 室内设计: HBA • Photography / 摄影: HBA • Photography / 摄影: The Alpina Gstaad •

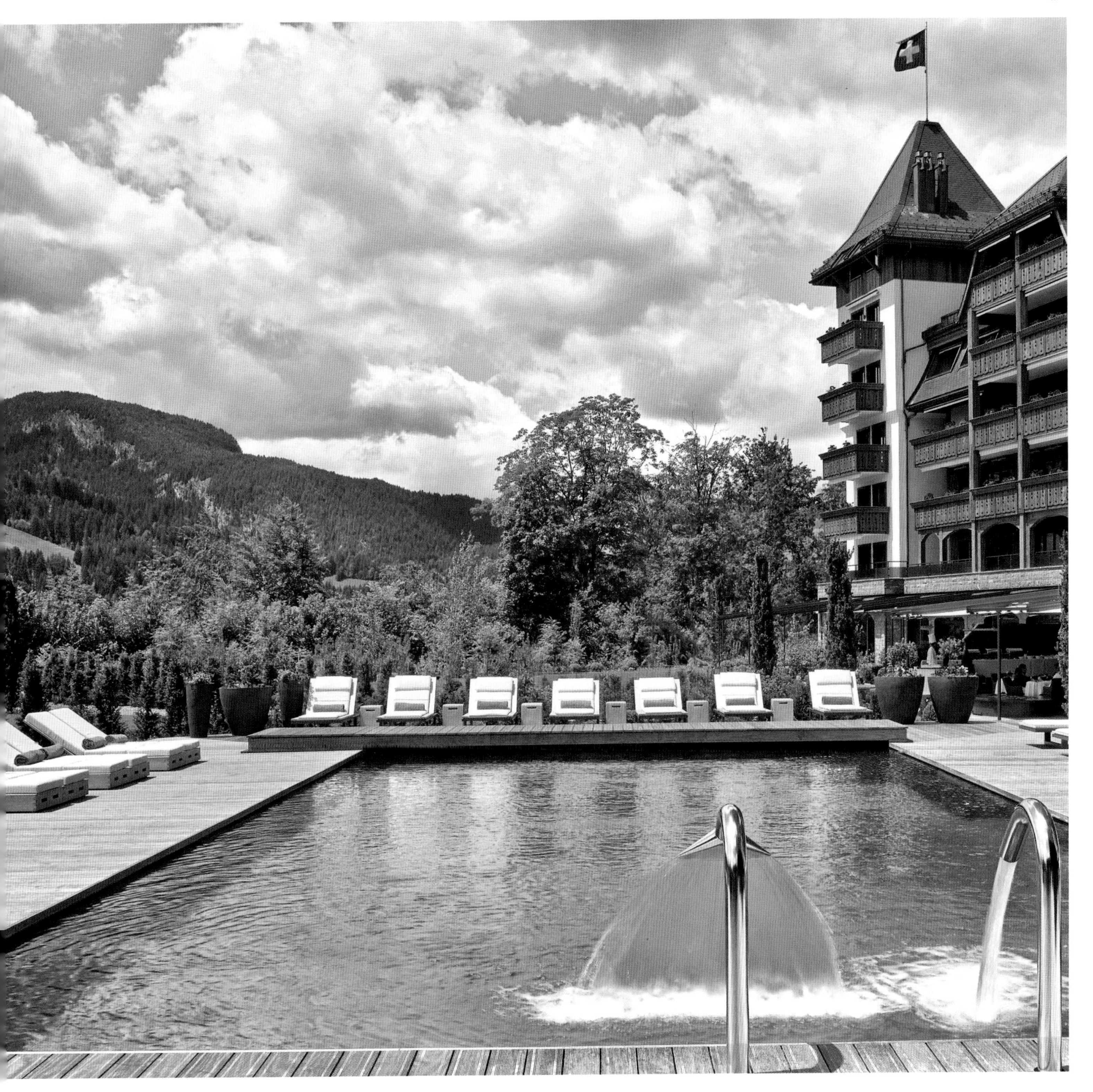

1 | 3
2 |

【细部解析】

1、吧台后板条木屏风的栅格设计使投射下来的灯光柔和而多变。

2、大堂两侧的立柱表面裹以压花皮革，使原本粗壮的柱子显得精巧雅致。

3、上下层空间以楼体贯穿起来，显得通透而气派。

MONTECRISTO
COHIBA
SIGLO VI

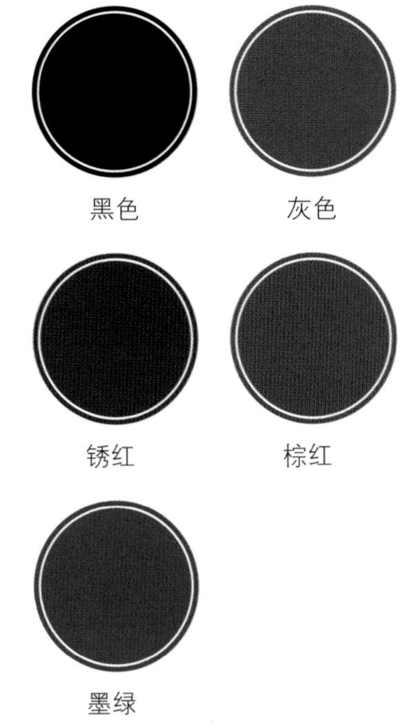

【主题色彩含义】：个性、沉静、内敛、低调奢华

壁炉周围的墙面由阿尔卑斯岩石建成，灰色的石材、地毯将锈红色的真皮沙发和同色系实木地板衬托得沉静而内敛。绿色和黑色的肌理挂画点缀其中，整体设计既传承了当地的极致工艺，亦体现出与众不同的低调奢华。

【细部解析】

酒店“Sommet”餐厅古朴的木天花充满当地特色，巨大的木横梁与其他公共区域的横梁一样，形成完美的燕尾造型。用于灯饰上的锻铁与用于沙发及座椅饰面的鞍皮随处可见。纯白的亚麻桌布、大量盆栽和鲜花装饰，都使餐厅整体环境显得十分清新雅致。

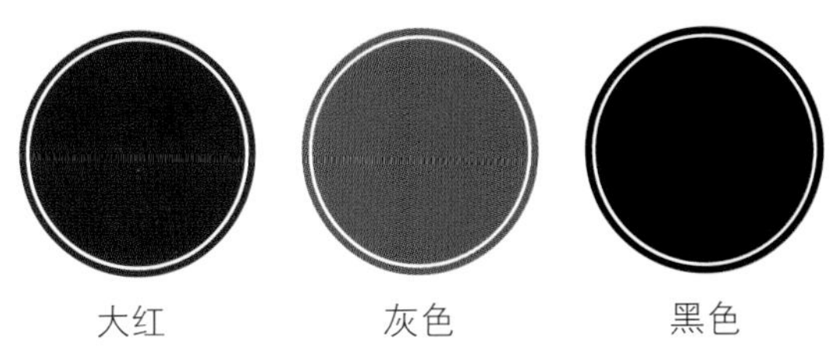

大红　　灰色　　黑色

【主题色彩含义】：时尚、华美、跃动

灰色与红色搭配在室内设计中很常见。纯度较高，但明度却不高的大红色，给人一种厚重而浓烈的感觉，也有一种热血沸腾般的强烈视觉冲击力，和灰色搭配运用可使红色不再浮躁，灰色不再单调，整体上显得时尚、跃动、华美。

1	3
2	

【细部解析】

1、木制天窗的造型十分别致、很好地解决了空间层次问题。

2、套房中的卫生间全部采用实木饰面装饰。

3、立柱表面裹以压花皮革，使粗壮的柱子显得精巧雅致。

LUXURY FOR HIM
WISH YOU WERE HERE
DECADE

KUNST

The Khyber Himalayan Resort & Spa, Gulmarg

开柏尔喜马拉雅度假酒店及水疗中心（印度）

开柏尔喜马拉雅度假酒店及水疗中心成立于2012年12月20日，该酒店是贡马地区首家豪华度假酒店，创办人为查漠和克什米尔地区的首席部长阿布杜拉先生。

开柏尔喜马拉雅度假酒店及水疗中心座落在海拔高达8825英尺的喜马拉雅山比尔本贾尔岭处，遍布七英尺的原始松树山谷。在此处游人可欣赏到令人叹为观止的Affarwat山峰白雪皑皑的壮观景象。酒店所处的得天独厚的地理位置和贡马地区独特的四季景象，使游客们总是在这里流连忘返。从贡马贡多拉可轻松步行到该度假酒店，这里有全世界最高的滑雪缆车。这部缆车每天承载大约600名游客和滑雪爱好者到达Affarwat山峰顶部的Kongdoori山脉，这座高达13780英尺的山脉是喜马拉雅山最高的滑雪地点。

这座国际一流的度假村主要由木材和石材建造而成。室内装饰采用了传统材料并且展示了克什米尔地区独特的工艺品。南姆达司的羊毛毡、当地编织的真丝地毯、带有刺绣装饰的家具、雕刻的胡桃木镶板，柚木地板以及纸质绘画等，无一不彰显着克什米尔地区丰富的手工业文化遗产。

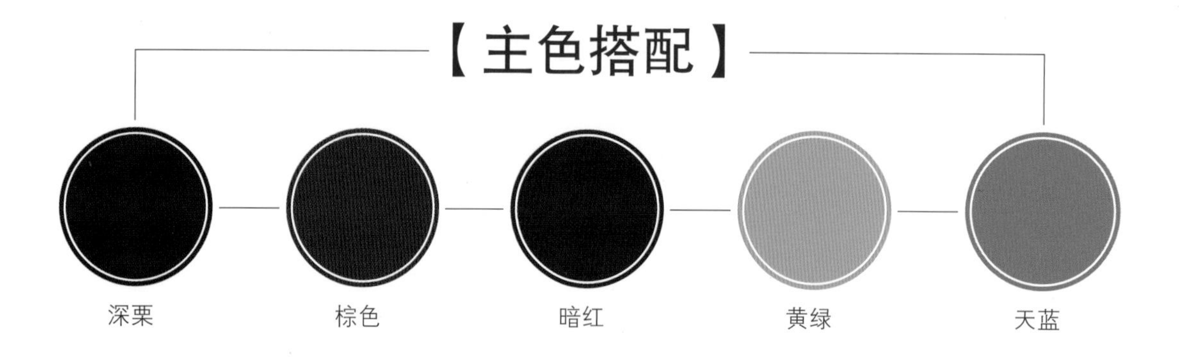

• Date of Completion / 竣工时间: 2013 • Interior Design / 室内设计: StudioB Architects • Photography / 摄影: Aditya Arya •

The Sunday Guardian
CONGRESS PREPARING

【细部解析】

酒店大堂的设计主要采用了传统材料，并且展示了克什米尔地区独特的工艺品。南姆达司的羊毛毡、当地编织的真丝地毯、精美的雕刻家具、胡桃木镶板等，无一不彰显着克什米尔地区丰富的手工业文化遗产。

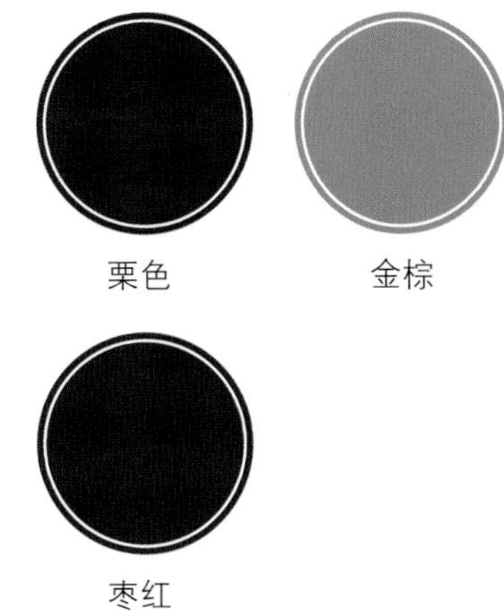

【主题色彩含义】：精致、优雅、庄重

枣红色是深红色的一种，既具有红色的热烈，又不失庄重感。与棕、栗色系搭配运用更会带给人优雅、精致、华美的感觉。

Waldorf Astoria Shanghai

上海华尔道夫酒店（中国）

上海外滩华尔道夫酒店由两栋大楼组成，一栋新落成的现代化塔楼连接着一栋全套房的新古典式历史建筑。曾经是具有传奇色彩的上海夜总会，也是上海遗留下来的为数不多的优秀建筑之一，始建于1911年。希尔顿集团委托HBA公司重新设计并复原这栋宏伟建筑，要求设计一个符合它称号的室内环境。这也是华尔道夫酒店建立的初衷，HBA要开创一种全新的能回答“全新的华尔道夫酒店应该是什么样子”的设计语言。

复原或改变历史遗迹是主要的设计挑战。因为原上海酒吧的流行性和照片存档都很大，影响了在这栋古建筑身上的创新，如新古典主义的室内、英国殖民风格的家具、中式基调的照明和手工艺品在人们心中烙印清晰。历史遗迹复原项目，复原要求极其注意细节的灵活性，这成了创新点。当设计师们到达现场时，发现了一个被他们忽略的圆柱，因此必须调整设计，这成为一个难题。与建筑相关联的一切，从天花板到墙体镶板，再到西西里岛大理石柱和伯明翰进口的彩色玻璃，都需要修复或再造。

最终，酒店完美结合了上海外滩闻名遐迩的历史文化与21世纪的繁华。在两幢综合大楼里，设置了260间高端配置的客房和套房，环境现代时尚的餐厅和酒吧，可俯瞰上海城市美景的精美华贵的宴会厅、豪华的水疗中心、设施完备的健身房、免费无线网络服务等，各功能设置精致巧妙，构成一幅完美的图画，展示出只属于上海外滩华尔道夫酒店的独一无二的宏伟和精致。

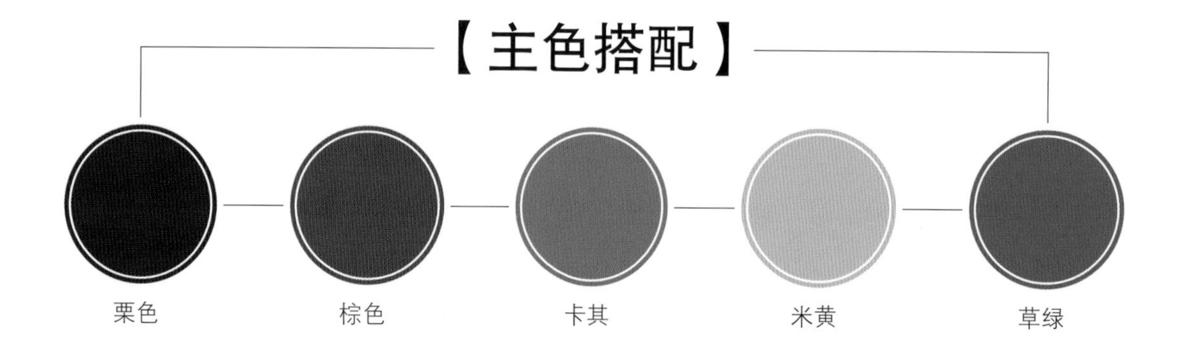

• Area / 占地面积: 1,500 m^2 • Date of Completion / 竣工时间: 2010 • Interior Design / 室内设计: HBA • Photography / 摄影: HBA • Client / 客户: Hilton •

【细部解析】

酒店建筑是典型的新古典主义设计风格。从天花板到墙体镶板，再到西西里岛大理石柱和伯明翰进口的彩色玻璃，这栋1911年建成的新古典式建筑及室内部分在档案图片和历史记录的帮助下已被精心地复原，并配备了现代奢华设施。

1

【细部解析】

1、独具匠心的冰冻展示酒库。

2、已有百年历史的廊吧内装有34米长的大理石吧台。

3、金棕色的欧式窗帘搭配香槟色的桌布和椅套打造出华丽的就餐环境。

1	2	3

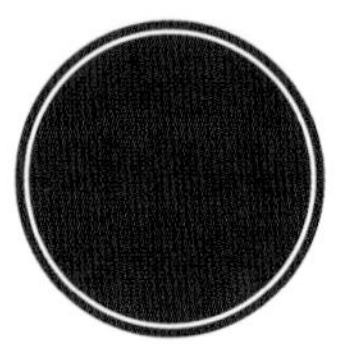

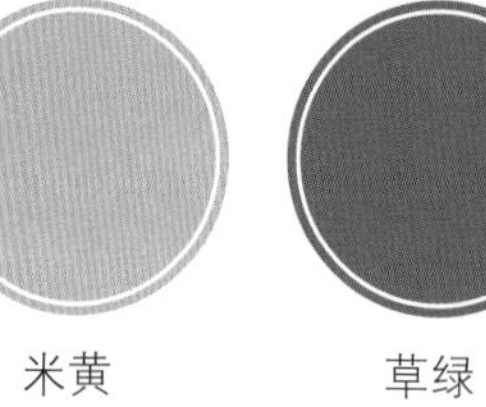

栗色　棕色　卡其　米黄　草绿

【主题色彩含义】：生动、温润、雅致

棕色系几乎是古典主义设计不可或缺的颜色，带给人庄重、大气、高贵的感觉。淡黄色的鲜花和青翠的绿色盆栽为古典格调的房间增添了一抹生机勃勃的气息。

Hotel Hospes Maricel

玛丽瑟尔水疗酒店（西班牙）

玛丽瑟尔水疗酒店是由16、17世纪保存完善的私家宅邸与现代化生态理念建筑相依而存的结构。在修缮完好的豪宅基础上，酒店委托Hospes团队采用中性色调的自然原石、实木及其他天然材质设计了极简主义风格的裸露石墙现代附楼建筑。

酒店倡导“可持续舒适”的理念，在海天相际之处为你还原壮丽海景与私密安然的默契交融。经过自然雕琢，玛丽瑟尔酒店拥有极具魅力的天然洞穴，经过现代精致的“Mar jades”建筑风格设计，为宾客打造一处舒适的休闲度假胜地。原石、实木、自然材质的地中海色调萦绕期间，为宾客带来真实的惊艳感受。

玛丽瑟尔酒店的所有空间设计无一不展现着艺术美感，为宾客带来视觉盛宴。宾客可以在此尽情欣赏无限的美景，感受沁人心脾的芳香，聆听海浪的声音，充分体会自然的味道、天然的雕饰。

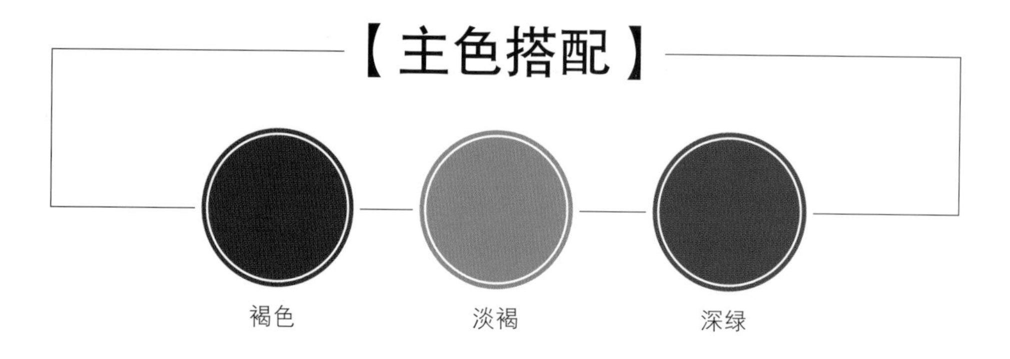

• Area / 占地面积: 100,000 m^2 • Date of Completion / 竣工时间: 2011 • Interior Design / 室内设计: Hospes Design Team •
Photography / 摄影: Hospes Design Team • Client / 客户: Evans Harch Pty Ltd and Education Queensland •

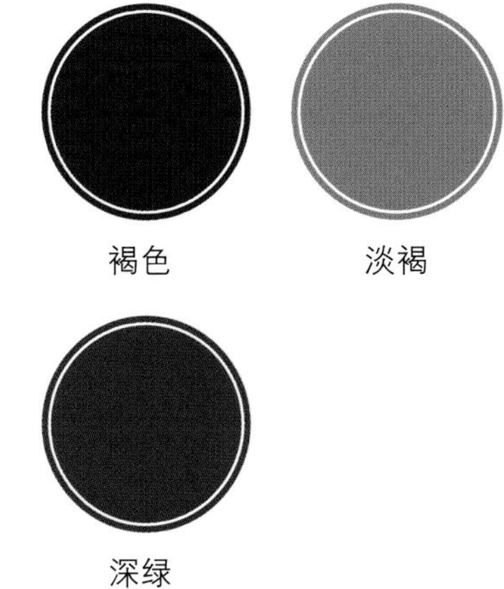

【主题色彩含义】：自然、优雅、舒适

褐色和深绿色都是自然界中常见的颜色，比如树木、昆虫等，饱和度偏低。将它们结合在一起运用，是一种非常自然、舒适、和谐的配色方法。

Hotel Principe di Savoia
普林西皮狄萨沃亚酒店（意大利）

普林西皮狄萨沃亚酒店于时尚之都米兰，始建于1927年，最近一次翻修设计于2009年，由Thierry W Despont、CDA Design、Francesca Basu几家团队合作完成，将其打造成为一家传奇酒店。

酒店设计理念是在尊重历史的同时革新并营造舒适的感受。为确保酒店是集传统与内涵为一体的21世纪休闲度假胜地，Thierry W Despont设计工作室尊重酒店的基本建筑特色，避免过多的修饰及粗制的装饰，将现代意大利风格的设计与古典灵感相融合。

在酒店入口、大厅、酒吧的设计中，Thierry W Despont设计工作室所面对的主要挑战就是：在尊重历史及Principe di Savoia酒店传统的同时，为宾客营造兴奋及复活感的体验。翻修后的酒店以更加奢华、高雅的格调为宾客营造如家放松的氛围。套房设计舒适豪华，有几种不同的风格，包括威尼斯、佛罗伦萨、新古典主义。设计中大量采用了大理石、镀金和锦缎等高档装饰材料和工艺。

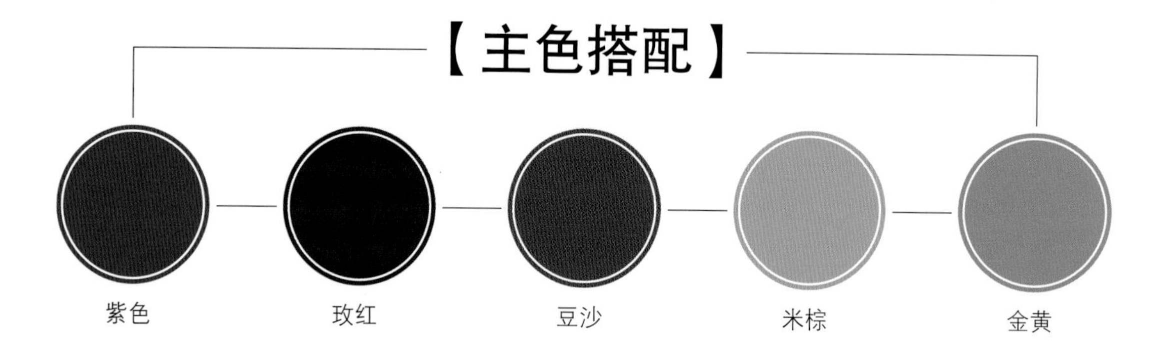

• Area / 占地面积: 1,305 m^2 • Date of Completion / 竣工时间: 2009 • Architecture Design / 建筑设计: CDA Design, The Office of Thierry W Despont, Ltd. • Interior Design / 室内设计: Francesca Basu Designs Limited, Celeste Dell'Anna Design, The Office of Thierry W Despont, Ltd. • Photography / 摄影: Hotel Principe di Savoia • Client / 客户: Hotel Principe di Savoia •

HOTEL PRINCIPE DI SAVOIA

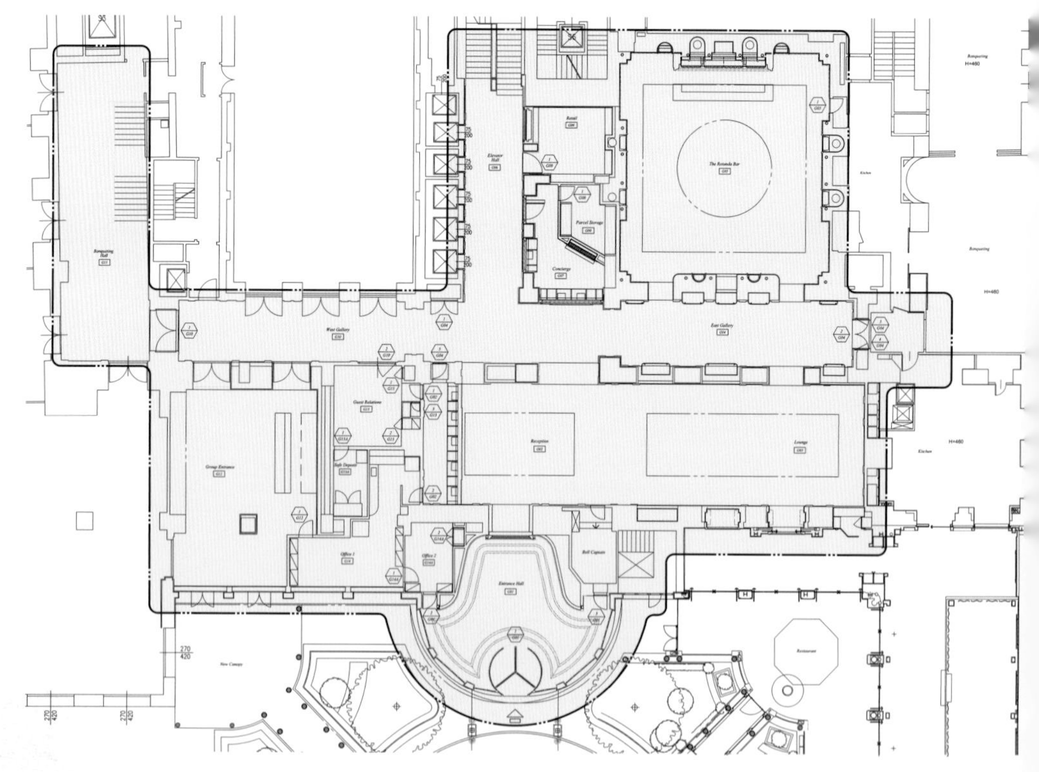
Retail
Elevator Hall
The Rotunda Bar
Parcel Storage
Concierge
West Gallery
East Gallery
Guest Relations
Group Entrance
Safe Deposit
Reception
Lounge
Office 1
Office 2
Bell Captain

【细部解析】

Acanto餐厅的设计体现了现代意大利风格与古典灵感的巧妙融合。极具才华的主厨Fabrizio Cadei以他娴熟的手法和创意的厨艺，为食客烹饪出了既保留意大利传统又别致的新派意式佳肴。

Manon

【细部解析】

吧台是空间的中心，有色玻璃的装饰直接受到Murano传统的启发。

黑色　玫红　棕色　金色　草绿

【主题色彩含义】：华美、富贵、丰茂

金黄色的墙纸、靠垫、床品和玫瑰红的窗帘、床幔、沙发，以及翠绿色的装饰品大胆地组合运用，在黑色和棕色的衬托下显得雍容华贵。

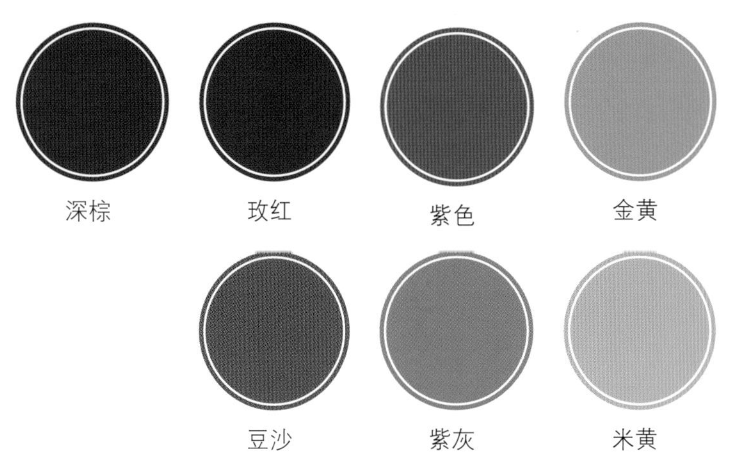

【主题色彩含义】：高贵、璀璨、雅致

金黄色与紫色、玫瑰红三种原本冲突的色彩在低饱和度的米黄、紫灰、豆沙色的调和中显得协调而富于变化。

【细部解析】

总统套房配备三个华丽的卧室、起居室、餐厅和一个私人游泳池，从家具到陈设配件，再到装饰材料和布艺面料，每个细节的设计和陈设装饰都十分精致考究。

collect

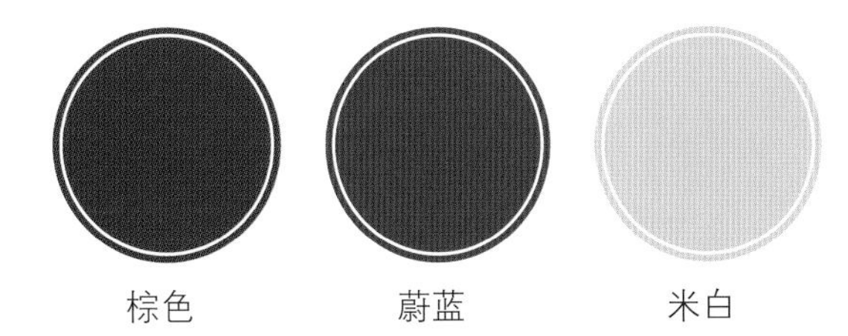

棕色　　蔚蓝　　米白

【主题色彩含义】：地中海、自然、明朗

蓝白相间的色彩搭配，结合古典花纹布艺装饰，成功地演绎了蔚蓝色的浪漫情怀和海天一色的纯美自然感觉。也很好地体现了地中海风格的设计灵魂。

Hotel Le Meurice

莫里斯酒店（法国）

莫里斯酒店是巴黎六家顶级酒店之一，从19世纪开始就被人们冠以“国王酒店”之名。莫里斯酒店是极具创造力的宫殿酒店，设计师将充满艺术品味的生活置于酒店历史中，将酒店的历史与法国的历史、巴黎的壮丽与世界的渴望相融合。

2007年莫里斯酒店主人Franka Holtmann委托Philippe Starck修复酒店，重塑其富饶、惊艳的动人形象。法国设计师的回应是：充分考虑色彩与光线，并对家具进行全新诠释，传达出透明及动感，打造“隐形的翻新”。与此同时Ara Starck原创了一幅油画作品，并将其永久陈列在乐·达利（Le Dalí）餐厅，向酒店最有象征意义的宾客萨尔瓦多·达利（Salvador Dalí）表示敬意。在莫里斯酒店丰富舒适之感不仅局限于酒店的公共区域。

在2008年12月，Holtmann委托Charles Jouffre为宾客打造崭新的、更温暖的氛围，具体细节包括奢华的窗帘、卡尼尔剧院（Garnier Opera）会堂前厅的帷幔等饰物。作为一位精益求精的工匠及艺术家，Charles Jouffre将宫殿酒店融入全新特色，旨在打造独树一帜的宫殿式酒店。酒店内160间宽敞、布置华美的套房都具有18世纪时期高贵的皇家韵味。

如今，莫里斯酒店是融合了过去与现在，集幽默与魅力于一身的宫殿式奢华酒店。

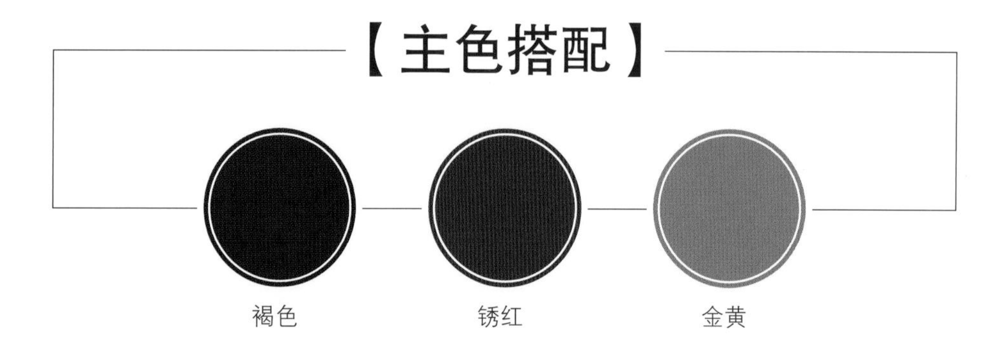

• Area / 占地面积: 27,000 m² / • Interior Design / 室内设计: Philippe Starck & Charles Jouffre •
Photography / 摄影: Le Meurice • Client / 客户: Le Meurice •

【细部解析】

莫里斯酒店大堂的设计别具匠心，尤其是这个既现代又古典的“壁炉”前所未有：燃烧的蜡烛和镜子是传统壁炉中“光亮”和“温暖”这两种含义的意象传达。

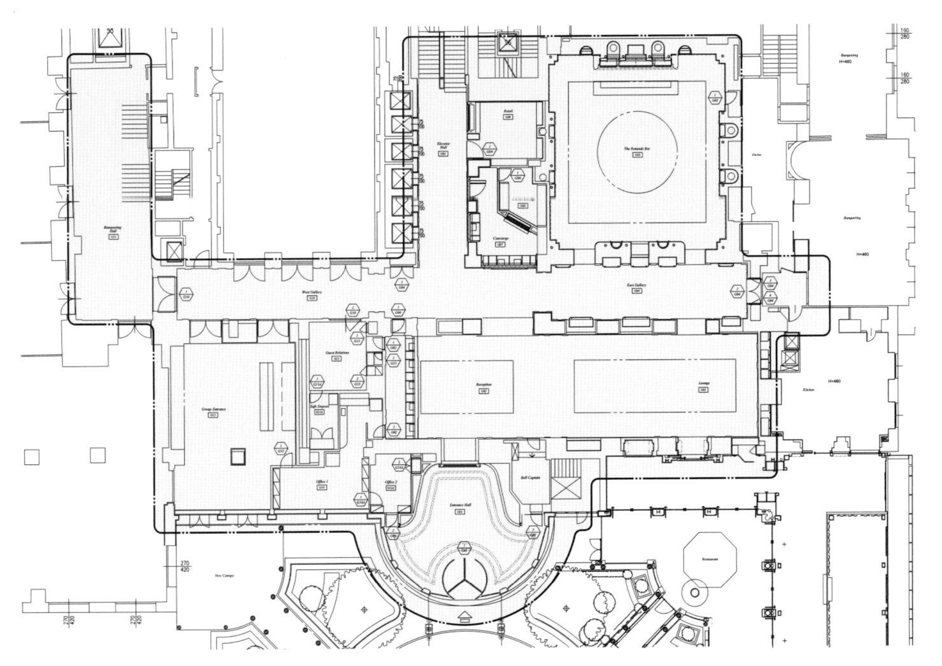

【细部解析】

乐・达利餐厅是大堂的开放式餐厅。从天花的巨幅油画到桌椅台灯，创意全都出自20世纪天才艺术家达利的画作、雕塑和手稿。

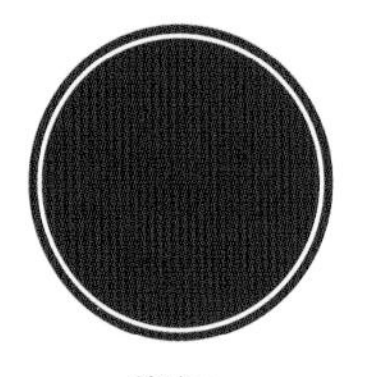

金棕　　浅棕　　米灰　　豆沙　　紫红

【主题色彩含义】：法式、浪漫、古典

金色、棕色系是古典设计常用的基础色，但单独使用时很难展现出一种个性的视觉效果，而把它们与粉色、红色、紫色等颜色大胆地搭配运用时，会实现一种生动、新鲜、个性的“后古典主义”风采。

The Dorchester

多切斯特酒店（英国）

多切斯特酒店于1931年对外开放，仅在数月之内，便以奢华的装饰及无可匹敌的优质服务赢得了世界顶级酒店地位的美誉。

酒店每一层都配有不同花色的手工地毯，设计师在Orchid套房的改造设计中，采用蓝白相间的配色方案和别出心裁的石膏装饰，以及一盏六脚枝形吊灯和中式橱柜，将洛可可艺术风格进行了充分的演绎。传奇的舞台设计师Oliver Messel设计了酒店一些最为惊艳的房间，包括Oliver Messel套房及屋顶套房。

多切斯特酒店的Spa经获奖的Fox Linton团队彻底修复后于2009年对外开放，重新设计的两层水晶套房为宾客提供了体验伦敦奢华的好去处。Spa以九间理疗室为特色，包括两间双人套房、mani-pedi套房，及整洁的放松室。

纽约室内设计师Alexandra Champalimaud彻底刷新酒店的三个著名套房，The Audley、Terrace及Harlequin套房，在2007年被共同叫做“屋顶套房”。2012年多切斯特酒店推出由Alexandra设计翻新的22间崭新套房。全新的套房设计在诠释明快现代的设计元素的同时展现传统的英伦之美。

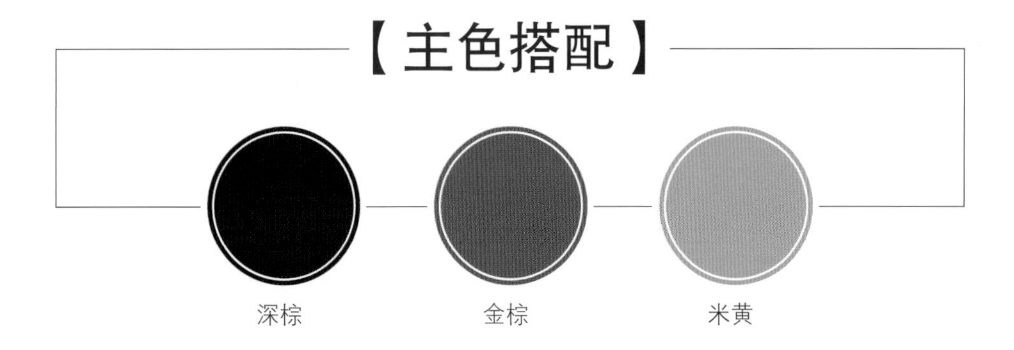

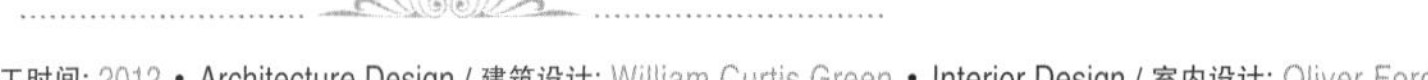

• Area / 占地面积: 37,161 m^2 • Date of Completion / 竣工时间: 2012 • Architecture Design / 建筑设计: William Curtis Green • Interior Design / 室内设计: Oliver Ford, Oliver Messel, Fox Linton Associates, Alexandra Champalimaud • Photography / 摄影: The Dorchester • Client / 客户: The Dorchester •

【细部解析】

光亮的米黄大理石镶嵌深色花纹，与墙上的镜面、金黄色的壁画、花纹和立柱交相辉映，在暖色灯光的映衬下，整个空间显得光彩夺目、熠熠生辉。

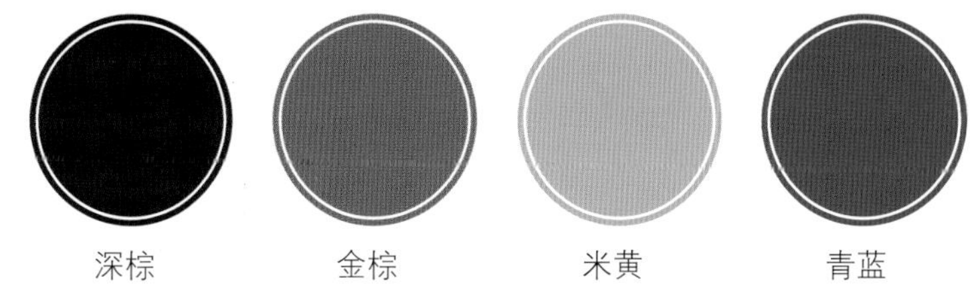

【主题色彩含义】：洛可可、雅致、清朗

米黄色墙面与金棕色沙发为房间明确了色彩基调，而深综色的中式橱柜与西式古典风格的设计巧妙地融为一体。青蓝色的窗帘与青花瓷器形成色彩上的呼应，也与背景色形成生动的对比。

Hotel Plaza Athenee

雅典娜广场酒店（法国）

雅典娜广场酒店占据优越的地理位置，位于法国巴黎香榭丽舍大道与埃菲尔铁塔之间，是巴黎时尚、娱乐与商务的中心地带。

奢华酒店具有华丽的巴黎黑色薄花呢装饰，设有191间客房，包括45间套房，宾客可以欣赏到埃菲尔铁塔西部、别致的蒙田大街或魅力庭院的景色。

位于雅典娜广场酒店第五层的皇家套房最近由设计师玛丽·何塞（Marie-José Pommereau）设计翻新。皇家套房总体面积450m^2，堪称巴黎最大的套房。设计师在保留房间陈设原有古典特色的同时，将套房以更加现代的外观展现在宾客面前。来自路易十五及路易十六时期的家具，画作及古董反应了纯粹的法式室内设计传统。同时，室内色彩方案也做了调整：第一间休息室充满紫色的光影，而第二间则强调金色与粉红色的对比。酒店为男宾客提供杏色调客房，为女宾客提供装饰为淡紫色的客房。

酒店自1911年开放以来，吸引了无数追求高雅生活的巴黎及国际宾客。基于“红色”的暖色调为酒店增添了魅力。酒店由内而外，从建筑物到服务，都将传统元素体现得淋漓尽致。

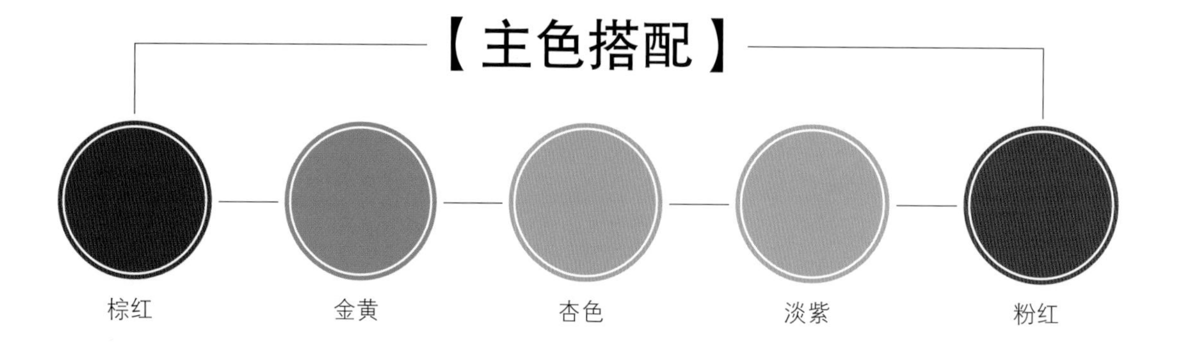

• Date of Completion / 竣工时间: 2012 • Interior Design / 室内设计: Marie-José Pommereau • Photography / 摄影: Hôtel Plaza Athénée • Client / 客户: Hôtel Plaza Athénée •

1	2	3

【细部解析】

1、红色阳伞和翠绿的叶饰构成醒目的酒店外观。

2、酒店大堂的理石地面上铺有厚厚的洛可可花纹地毯。

3、法国古典风格的家具、壁画和装饰营造出宫廷般的奢华感。

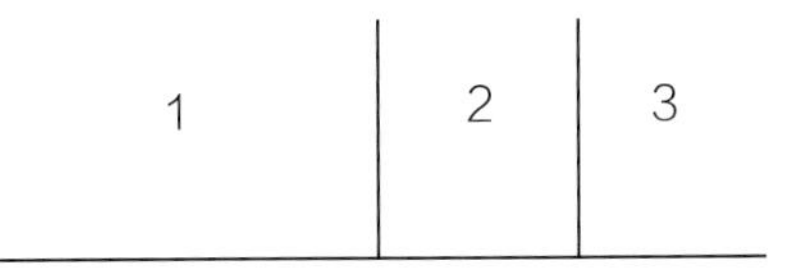

【细部解析】

1、哥伯林家饰艺廊(La Galerie des Gobelins) 提供下午茶服务。

2、哥伯林家饰艺廊采用统一协调的棕色系打造古典与现代相融合的格调。

3、做工精良的座椅细节与马赛克地砖细部展示。

CD

棕红

金黄

杏色

淡紫

粉红

【主题色彩含义】：浪漫、宫廷、奢华

在皇家套房的设计中，来自路易十五及路易十六时期的家具、画作及古董反映了纯粹的法式室内设计传统。同时，设计师对室内色彩方案也做了调整：第一间休息室充满紫色的光影，而第二间则强调金色与粉红色的对比。

The Beverly Hills Hotel and Bungalows

贝弗利山庄别墅酒店（美国）

贝弗利山庄别墅酒店也被称为“粉红宫”，位于贝弗利山的历史文化区。酒店共设有208间客房及套房，包括23栋独一无二的平层别墅。酒店的花园苍翠繁茂芳香扑鼻，游泳池边棕榈树成行，漂亮的客房为宾客带来舒适体验，公共区域令人联想到好莱坞永恒的魅力，每一处设计都为宾客成功地营造出放松的度假氛围。

经典的淡粉色墙及漂亮的花园景观后是每一间总统套房极其华美的环境。具有瀑布景观的游泳池及水下扬声器放在露台上，半圆形休息区放置躺椅及太阳椅、桌子可以围坐10人，壁炉、健身区也设于此。

两间崭新的总统套房是对传统设计的现代诠释，补充并更新具有历史意义的地中海特色。每一间房间的陈设都是独特的，集不同的式样、面料、色彩、窗帘、质地于一身。室内的餐饮区将南加州风情与自然相融合，压制玻璃的枝形吊灯为餐饮区照明，矩形木桌由定制的皮革椅环绕，可环坐10人。

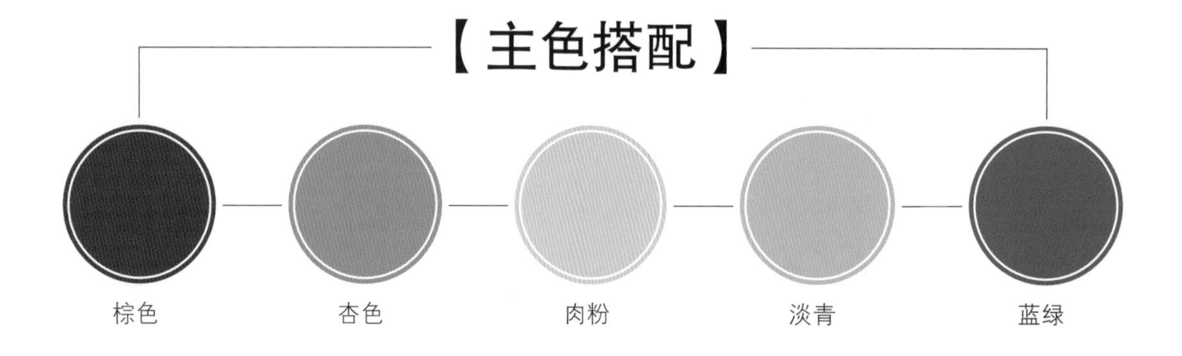

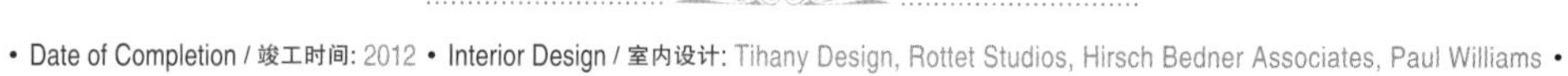

• Date of Completion / 竣工时间: 2012 • Interior Design / 室内设计: Tihany Design, Rottet Studios, Hirsch Bedner Associates, Paul Williams •
Photography / 摄影: The Beverly Hills Hotel and Bungalows • Client / 客户: The Beverly Hills Hotel and Bungalows •

【细部解析】

将整个酒店都点缀上蓝绿色，是设计师赋予酒店的巧妙创意之一。蓝绿色与白色相间条纹美观而醒目，和同样颜色的香蕉叶图案的壁纸，这种设计既可以最大限度地保持酒店的原貌，又可以不动声色地使其成为该酒店的标志性特征。

棕色

杏色

肉粉

淡青

蓝绿

【主题色彩含义】：维美、柔和、清新

本案的套房设计中采用了蓝绿色、肉粉色、杏色、棕色等几种贝弗利山庄的独特色彩。蓝绿色与棕色之间由提高明度、降低饱和度的淡青绿色、肉粉色和杏色形成巧妙而柔和的过渡。几种颜色组合在一起运用，既有冷暖上的对比又无比自然、丰富和协调。

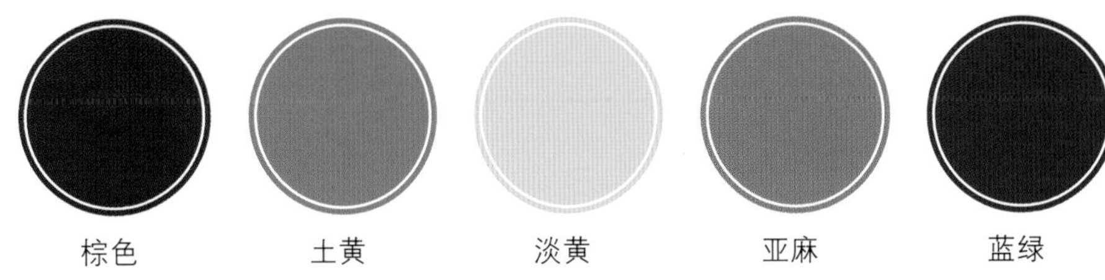

【主题色彩含义】：温暖、璀璨、金碧辉煌

棕色系和米色系构成的色彩背景柔和而自然，淡黄色的台灯、花卉与同样颜色的毛毯形成呼应的同时，也为背景增添了一抹醒目的光辉。蓝绿色是酒店的主题色之一，在这里与其他色彩形成了鲜明的对比。

KONKANI

1 | 3 | 5 | 7
2 | 4 | 6 |

【细部解析】

1、套房卧室采用的米色系床上用品温馨雅致。

2、套房内的沙发椅、落地灯和花瓶构成完美的搭配。

3、倒置酒杯形状的茶几精巧别致。

4、胡桃木制成的简洁写字台和镜框形成呼应。

5、套房内的步入式衣橱。

6、浴室外的梳妆台和造型独特的座椅。

7、浴室采用统一的白色调衬托室内的绿植装饰和门外的自然景观。

Hotel de Crillon

克里伦酒店（法国）

克里伦酒店是全球最古老的豪华酒店之一，始建于1785年，由路易十五（Louis XV）委任当时著名的建筑师Jacques-Ange Gabriel负责设计。酒店位于巴黎香榭丽舍大道一端，地处协和广场北面。

这栋房子曾一度租予奥蒙公爵（Duke d' Aumont），然后才转由Crillon家族持有所有权，直至1907年。1909年，Hotel de Crillon为酒店辉煌时代揭开序幕。时至今日，Hotel de Crillon仍遗留着过去华丽时光的痕迹，比如Salon des Aigles天花板上的Wedgewood圆形陶瓷浮雕和Les Ambassadeurs餐厅的水晶吊灯与大理石地板。

这家五星级酒店位于两座完全相同的建筑内，共有147间客房，包括44间套房，均以路易15世（Louis XV）风格装修。在酒店的任何一间套房都可以欣赏到180度的巴黎美景。酒店是法国唯一拥有大都会博物馆木艺品展出的奢华酒店。历史悠久的沙龙经常用做举办各类外交活动的场地，并在1919年荣幸的举办了国际联盟的签定仪式。

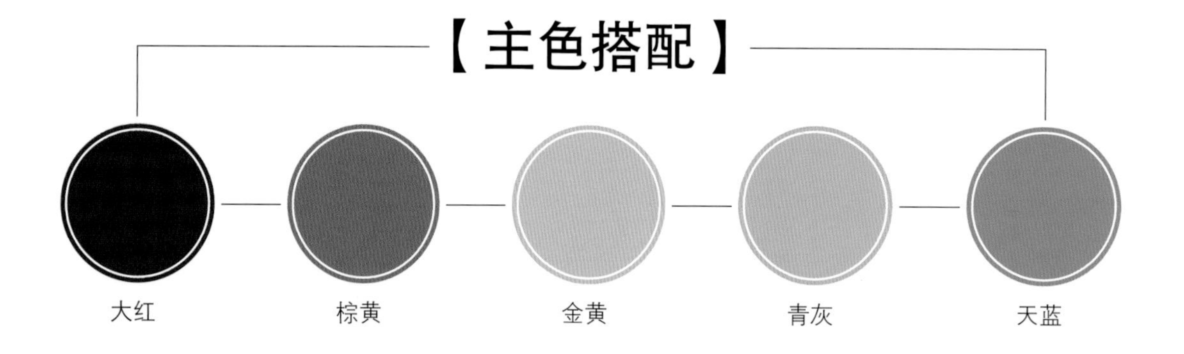

• Area / 占地面积: 14,000 m^2 • Date of Completion / 竣工时间: 2011 • Architecture Design / 建筑设计: Jacques-Ange Gabriel • Interior Design / 室内设计: Jacques-Ange Gabriel • Photography / 摄影: Eric Cuvillier •

【细部解析】

在一座宏伟的外墙后面，有一处由当时最优秀的艺术家和工匠共同完成的奢华私人住宅。这就是克里雍大饭店的起源处，它的建造是用来欢迎世界上最伟大的使者。

【细部解析】

绿色藤类植物攀爬在古老的建筑外墙上，为酒店中庭带来生机勃勃的气息，室外宽大的座椅和白色的阳伞、桌布、花束及椅垫在绿色的衬托中显得明亮洁净，营造清新愉悦的休闲就餐环境。

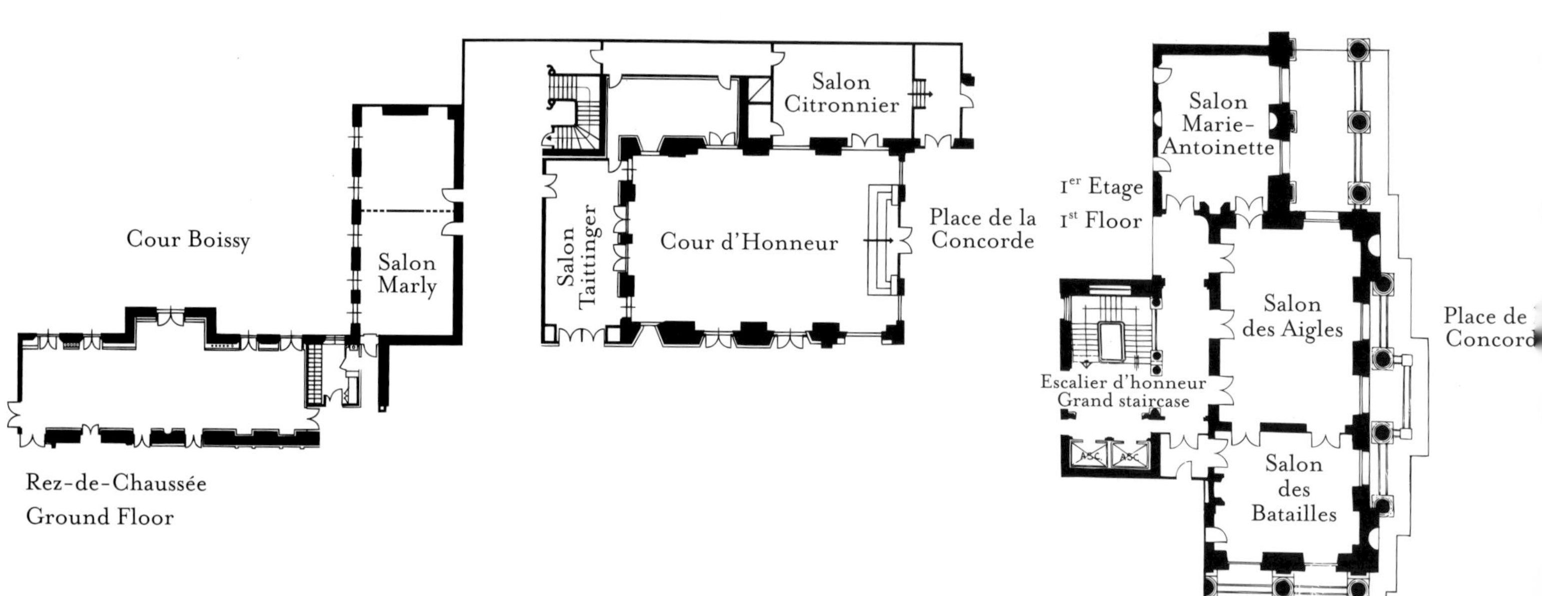
Salon Citronnier
Salon Marie-Antoinette
Ier Etage
Ist Floor
Place de la Concorde
Cour Boissy
Salon Marly
Salon Taittinger
Cour d'Honneur
Salon des Aigles
Place de
Concord
Escalier d'honneur
Grand staircase
ASC
ASC
Salon des Batailles
Rez-de-Chaussée
Ground Floor

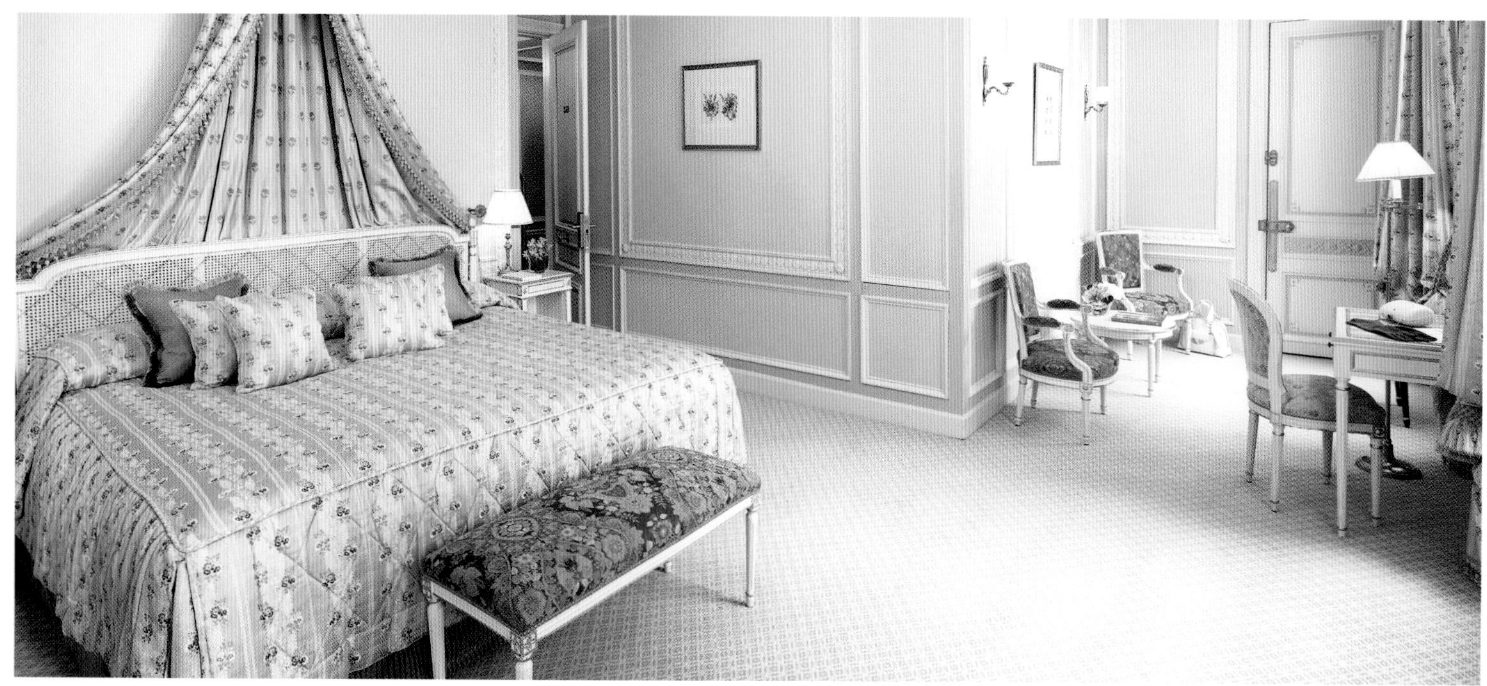

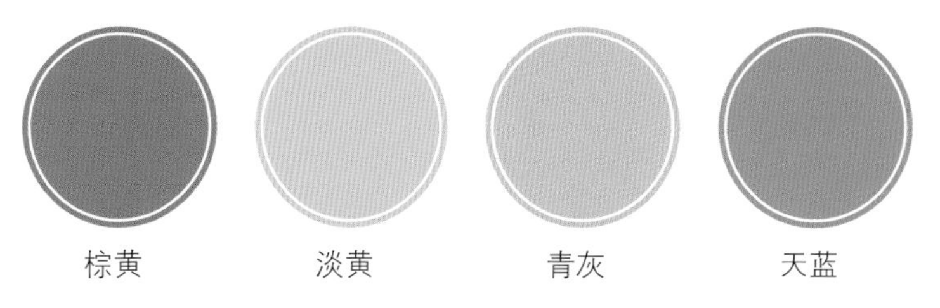

【主题色彩含义】：悠然、恬静、明亮

天蓝色明度和饱和度适中，与明亮的黄色搭配运用，会让人联想到阳光、蓝天、晴朗等舒适、自然的感觉。结合棕色、杏色、米色和青灰色等饱和度偏低的颜色，可以让这两种对比色显得丰富、协调而优美。

Hotel Baltschug Kempinski Moscow

巴尔舒格凯宾斯基酒店（俄罗斯）

巴尔舒格凯宾斯基酒店是2012优质五星级酒店，同时作为莫斯科首家五星级酒店，迎来“表现和感觉”为主题的20周年品牌纪念。

2012年夏天酒店以奢华时尚的装修和久负盛名的服务准备好展现它升级后的容光。新餐厅有开放式厨房，还有为私人品鉴准备的酒室、现代化的设计、菜单上现代与古典风味兼备的菜品，都吸引宾客流连忘返。大厅里意大利大理石和水晶灯映在酒杯里的影子似乎都在欢迎客人。

修饰一新的克兰兹勒咖啡馆和大厅酒吧以及新会议室，仅仅是酒店全景的一部分。巴尔舒格凯宾斯基酒店总是将俄罗斯首都莫斯科灿烂的历史和著名景点相结合。酒店拥有一百多年的历史，在这里著名俄国艺术家创作的描绘巴尔舒格美丽风光的艺术作品，今天仍然是酒店客人最美好的回忆。

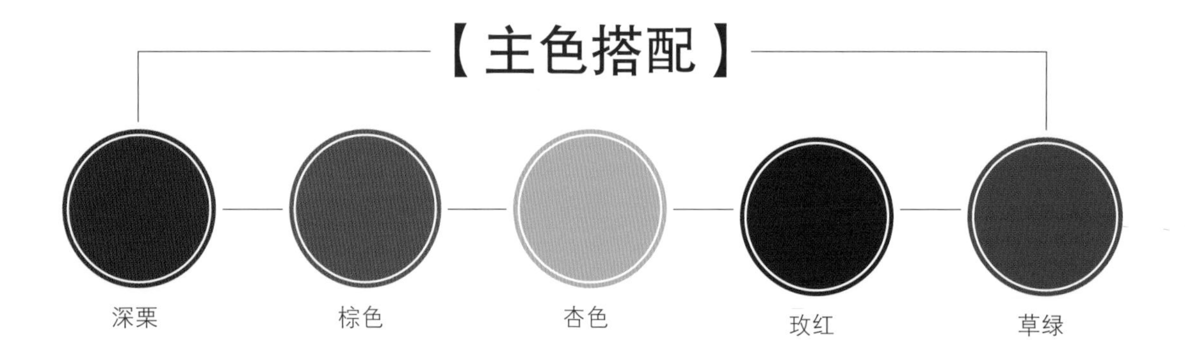

• Area / 占地面积: 14,000 m² • Date of Completion / 竣工时间: 2012 • Design Firm / 设计团队: CADÉ • Photography / 摄影: Kevin Kaminski •

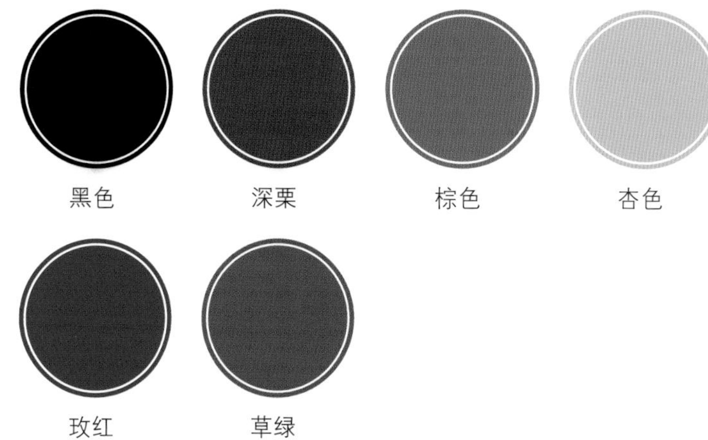

【主题色彩含义】：明快、典雅、自然

玫瑰红色透彻明晰，既包含着孕育生命的能量，又流露出含蓄的美感，华丽而不失典雅。而绿色给人玫瑰花叶的感觉，两者搭配运用十分协调自然，给人楚楚动人、高贵典雅的美感。

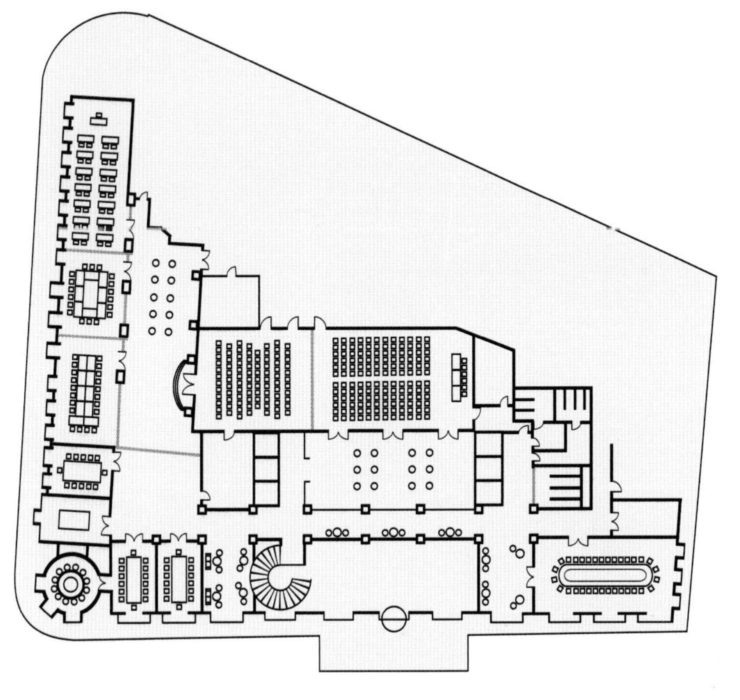

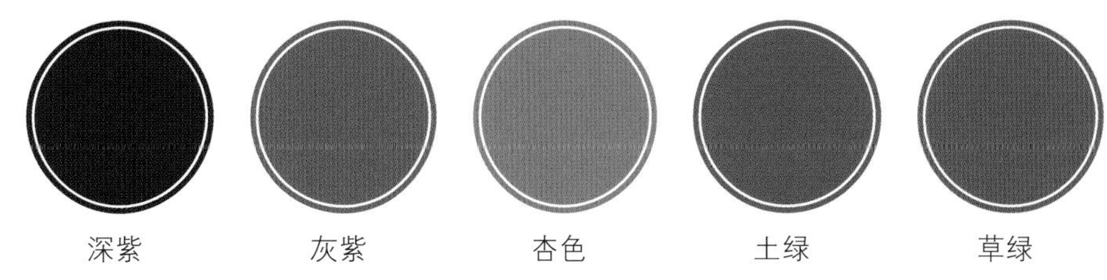

【主题色彩含义】：丰富、优雅、尊贵

紫色与绿色搭配属于二次色（间色）配色法的一种。由于紫色和绿色都含有蓝的成分，所以二者搭配起来比较容易协调。而以这两种颜色为基础，减低饱和度的两种色系搭配，产生的色彩效果更加丰富、柔和。

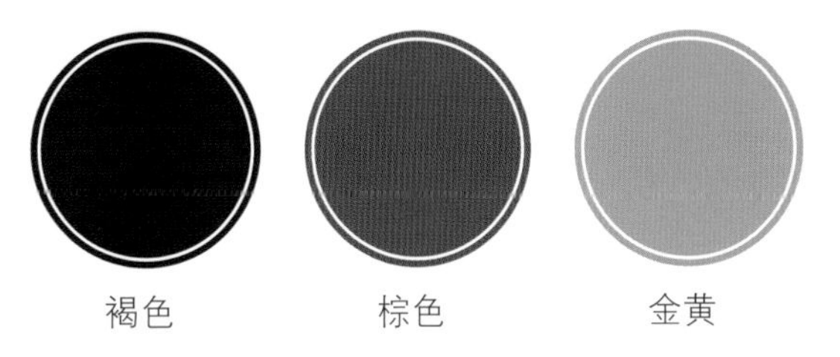

【主题色彩含义】：富丽堂皇、复古、光明

同色系的搭配很容易打造出协调舒适的感觉。褐色实木酒柜、吧台和座椅边框搭配棕色调豹纹图案地毯和金黄色椅子、立柱，一种富丽堂皇的古典韵味油然而生。

Kempinski Hotel & Residences Palm Jumeirah

迪拜棕榈岛凯宾斯基酒店及公寓（阿联酋）

棕榈岛凯宾斯基酒店坐落于阿联酋迪拜标志性海岛上，位置得天独厚。宫殿般华丽的休息场所使之成为一个名副其实的地标性建筑，这个如画般的度假天堂为宾客提供传统欧式奢华生活体验和无数永恒美好的回忆。

大地、海洋与奇思幻想相结合的棕榈岛凯宾斯基酒店欢迎来自世界各地的游人。在酒店内既可以俯瞰远处连成一线的环岛礁湖，也可以欣赏阿拉伯海静谧的海景。

该酒店有着这一地区最宽敞的住宿条件：244间雅致套房、112m^2到894m^2不等的高层公寓和别墅。多种类型的餐厅和酒吧为客人提供种类丰富的料理，包括布鲁耐罗意式餐厅、海滩烧烤酒吧和基韦斯特酒吧。其中布鲁耐罗餐厅装修考究，可以观赏室内花园的美景。平台上宽敞的法式门方便客人在宜人的天气享受户外进餐。基韦斯特酒吧主要采用华丽的皮革制品和实木装饰，以无瑕服务为高端业内人士准备了私人灯光和芳香的高级雪茄。

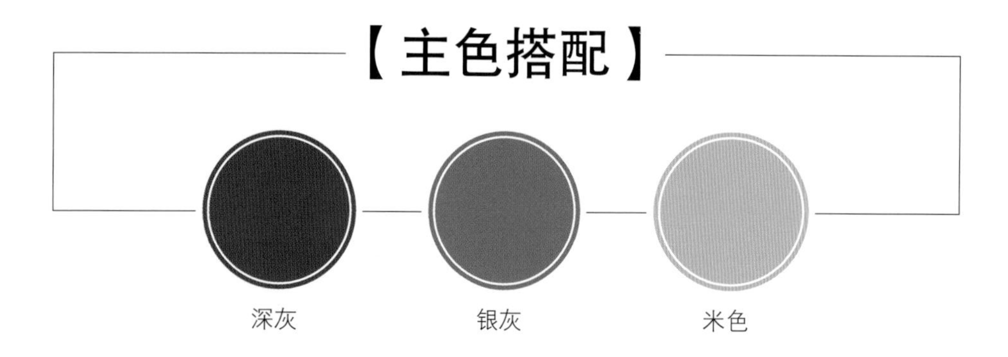

• Area / 占地面积: 14,000 m^2 • Design Firm / 设计团队: Godwin Austen Johnson Architects • Photography / 摄影: Kempinski Hotel & Residences Palm Jumeirah •
Client / 客户: Kempinski Hotel & Residences Palm Jumeirah •

【细部解析】

这是迪拜棕榈岛凯宾斯基酒店内具有三间卧室的顶层复式套房，每个套房都包括一个大露台或阳台以及一间设备齐全的厨房。起居室选用古典风格的家具和带有精美浮雕花纹的地毯，呈现出极致奢华的感觉。

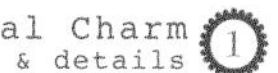

3 Bedroom Suite

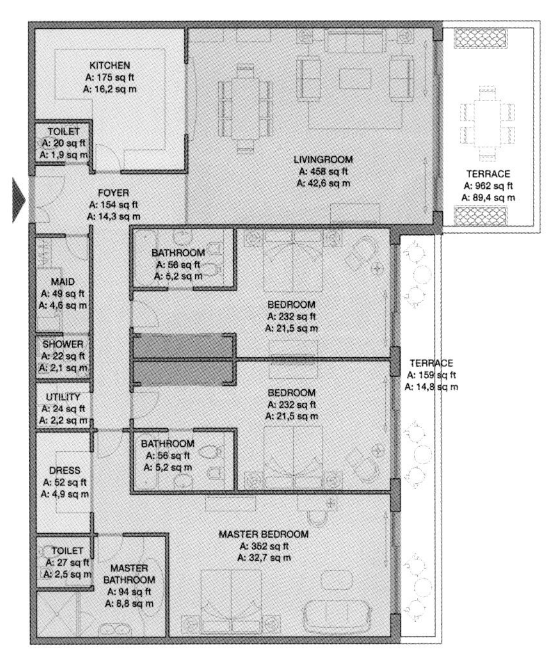

Total Area: 210-263 sq m

4 Bedroom Penthouse Suite

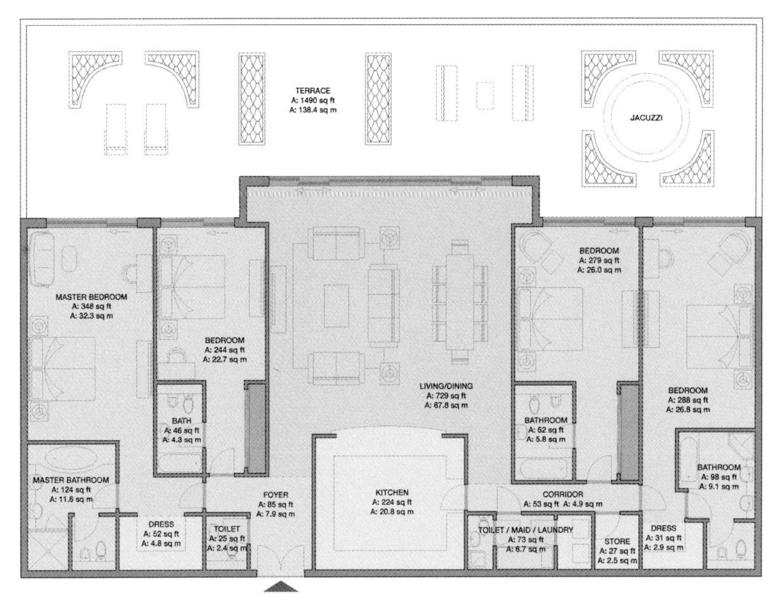

Total Area: Superior Penthouses 285-303 sq m, Deluxe Penthouses 366-492 sq m

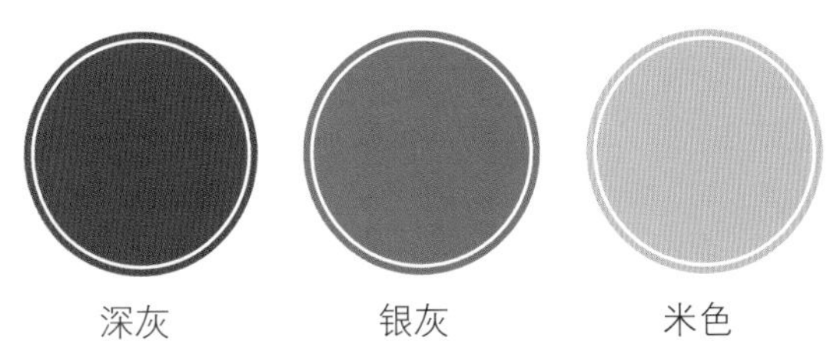

【主题色彩含义】：淡雅、高级、气质

灰色是介于黑和白之间的一系列颜色，按照色阶可以大致分为深灰、中灰和浅灰等。没有任何色彩倾向的灰色与任意颜色的搭配都可以协调自然。高明度的灰色或者银灰色搭配米色、杏色等中性色，可以营造出十分淡雅、润泽、柔美的氛围，是一种非常适合卧房的配色法。

Leon's Place In Rome
罗马里昂宫酒店（意大利）

意大利罗马里昂宫始建于19世纪的贵族宫殿中，如今是最现代的工业建筑酒店的代表。几百年前，该宫殿曾作为宗教活动地点，一走进入建筑就会注意到这个特点。大堂和公共区都有很重要的影响力。

宽敞空间和装饰的物件吸引着每一个进入到宏伟大堂的人。巨大的黑色天鹅绒装饰物在正中间的吊灯上摆动，吊灯由高级水晶和织物制成。这里几乎成为一个新哥特风格的缩影。威风的帐幔、精心设计的灯光、珍贵材料的漆面，都使古老宏伟的气氛成功复活。

里昂宫的56个房间和套房的主要基调为黑色、白色和珍珠灰。奢华、后现代的风格与古老的灵魂融合于宫殿之中。精致的设计与原有建筑和谐的融合在一起，例如，精致的天花板和阳台与新古典风格的铁制装饰物相搭配。地毯、华丽柔软的床头板、黑色家具和镜子通过准确的灯光映衬出来。房间配备齐全：LCD屏的有限电视、网线、电话独立空调、保险柜、小吧台和独立卫生间。从一些房间的阳台可看见古罗马城区。

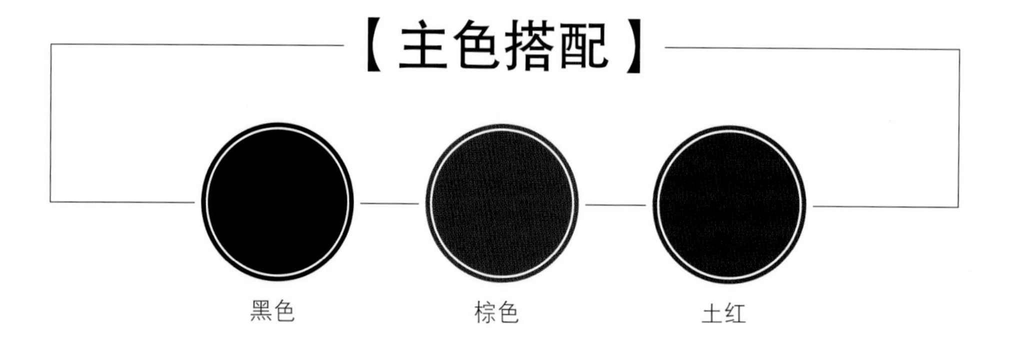

• Design Firm / 设计团队: Hotelphilosophy Creative Dept. in collaboration with Visionnaire by Ipe Cavalli • Photography / 摄影: Massimo Listri, Martina Barberini, Max Zambelli •

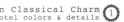

【主题色彩含义】：神秘、忧郁、性感

只用很少的颜色，有可能既表现出一座历经更迭的古老城市神秘而魅力的一面，又表现出它忧郁性感的另一面吗？位于罗马的里昂宫酒店成功做到了。大堂内沉稳内敛的棕色系，在明度反差巨大的黑白两色之间显得柔和而饱满。少量土红色点缀其中，更传达出忧郁、神秘，豪不浮夸的奢华感。

黑色　棕色　土红

Leon's place
hotel in rome

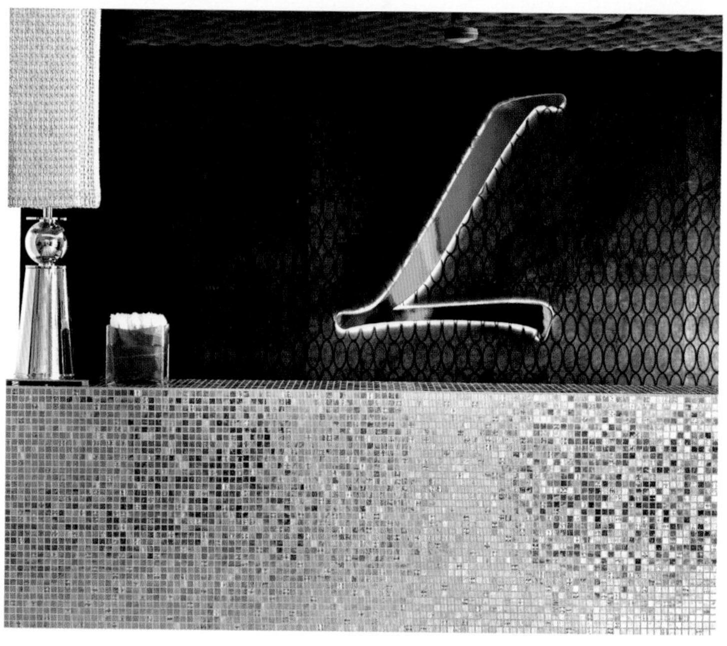

【细部解析】

1、黑色座椅皮面与光亮的黑色镜面形成材质对比。

2、水晶灯上装饰着黑色的羽毛新颖别致。

3、精致而低调的酒店标识。

4、网状背景上的光亮标识与镜面马赛克形成呼应。

5、镜面墙设计既可以提亮空间也可以拓宽视野。

6、棕色天鹅绒窗帘与黑色绒面沙发和地毯搭配，打造出忧郁内敛的奢华美感。

1	3
2	4

5	6

【细部解析】

Visionnaire咖啡吧再一次呈现出非凡的创造性视觉体验。吧台立面采用的银色瓷砖十分昂贵，优良的反光效果使它可以随着光线的改变而呈现出千变万化的色彩，与吧台上方的吊灯交相辉映。

【细部解析】

客房的设计以珍珠灰和黑白作为基调，精致柔和同时个性十足。奢华、后现代的风格与古老的灵魂融合于宫殿之中。精致的设计与原有建筑和谐融合在一起，例如精致的天花板和阳台与新古典风格的铁制装饰物相搭配。地毯、华丽柔软的床头板、黑色家具和镜子通过准确的灯光映衬出来。

Arch. Statilio Ubiali

Add: Via Marconi nr.36, 24040 Verdellino (BG), Italy
Tel: 0039-35-4820655; Fax: 0039-35-4820655
Email: studioubiali@alice.it

Statilio Ubiali was born in Verdellino (Bg), Italy, in 1961. He attended the Politecnico of Milan – Architecture School and graduated from the University in 1988. He attended Architecture Associate Office from 1989 to 2005 and has worked in Architecture Private Office from 2005.

Arch. Ubiali's activities cover several fields, with many important works: interior design of villas, palaces and offices in Italy, Russia, Austria, Switzerland and Middle East; restoration of 4 and 5 stars luxury Hotels in Italy (Grand Hotel des Iles Borromees, Regina Palace Hotel in Stresa-Lake Maggiore); furniture collections design for classic style furniture firms; yacht interior design; fitting-out of stands in fairs and boutiques in Italy; design of wellness centres and design and restoration of historic gardens and parks.

Architetturambiente

Add: Architetturambiente st. ass. Bosi e Soloni Piazza Paul Harris 4, 48122 Ravenna (Italy)
Tel: +39.0544.420303; Fax: +39.0544.426141
Web: www.architetturambiente.com

Architetturambiente is a studio focused in providing services in the areas that the designers of Architetturambiente most love: architecture , interior design and pure design in ceramic tiles field. Key elements of their studio is their strong passion, fueled by ongoing research in various areas of design, and therefore characterized by a significant technological knowledge. Using the most sophisticated and innovative software it is possible to realize the "right" design, giving consistency to customer and their ideas without skippping any step. But the main objective of their work is the satisfaction of the Client, which is the focus of all their projects. Architetturambiente can provide all the the types of necessary services, from the pure concept design to the final construction , without skipping any step.

Four Seasons Hotel des Bergues Geneva

Add: Four Seasons Hotel des Bergues Geneva 33, Quai des Bergues 1201 Geneva Switzerland
Tel: +41 (22) 908 70 00; Fax: +41 (22) 908 74 00
Web: www.fourseasons.com

On the shore of Lake Geneva, surrounded by the Jura Mountains and Swiss Alps, presides a landmark hotel that has helped shape the history of a continent. Now, 179 years after the debut of this grande dame, Four Seasons Hotel des Bergues Geneva is once again transforming with the addition of 18 new rooms and suites and redecoration of 21 others by esteemed French designer Pierre-Yves Rochon. Crowning the renovation will be a rooftop terrace with sweeping views of Mont Blanc and the lake's famous Jet d'Eau fountain, plus an all-new spa, fitness centre and year-round pool.

"Many of our guests see Four Seasons as their pied à terre in Geneva," says General Manager José Silva. "The Hotel's ambience is that of an elegant private residence, and our 250 staff – including a team of personal assistants – is there to ensure every need is anticipated, and every wish granted."

HBA

Add: Two Peachtree PointeSuite 7001555 Peachtree Street, NEAtlanta, GA 30309 USA
Email: HBAglobal@HBA.com
Web: www.hba.com

World-renowned as the "Number 1 Hospitality Design Firm" (Interior Design) and honored in 2012 by the Gold Key Awards, Hospitality Design, Perspective Awards, the Boutique Design Awards and the European Hotel Design Awards; HBA, unveils the world's most anticipated hotels, resorts, and spas.

Leading the hospitality interior design industry since 1965, HBA remains keenly attuned to the pulse of changing industry trends governed by today's sophisticated traveler. The company's international presence, depth of experience, and detailed industry knowledge enables them to identify interior design trends at their source, make definitive predictions about new directions and innovations, and influence design standards at a global level. HBA's ultimate objective is to add value, raise standards and enhance the brand of a project's owner and operator.

HBA creates the signature look of traditional luxury brands, independent contemporary boutiques, urban resort spas, world-class residences, restaurants, casinos, and cruise ships. With over 1,200 designers around the globe in 16 offices and a recent expansion in several locations in Asia, HBA is a true global company.

HBA's international presence, combined with its extensive knowledge of the interior design industry, has facilitated the ability to rewrite the language of design with each new project. HBA is based in Los Angeles, Atlanta, San

Francisco, London, Hong Kong, Beijing, Shanghai, Tokyo, Singapore, Melbourne, New Delhi, Dubai, Moscow, Istanbul, Bangkok and Manila.

Hotel Baltschug Kempinski Moscow

Add: Hotel Baltschug Kempinski Ul. Balchug 1, 115035 Moscow
Tel: +7 495 287 2000
Fax: +7 495 287 2002
Web: www.kempinski.com/en/moscow

Located in the heart of Moscow, the Hotel Baltschug Kempinski Moscow has an air of European timelessness and Russian spirit . Whatever your need, whenever you need it, their team are there to ensure you get the best from your visit.Hotel Baltschug Kempinski Moscow – quite simply, for when you want to make the most out of Moscow.

Hotel de Crillon

Add: Hotel De Crillon 10, Place de la Concorde – 75008 Paris France
Tel: +33 0144711500
Web: www.crillon.com

In the designer's architecture it is vital to maintain the original character of the Flagship of the Concorde Hotels & Resorts Group, the Hôtel de Crillon is a member of Leading Hotels of the World.

After the French Revolution, for which the square was a central stage, the First Consul decided to remove the statue of Liberty which had taken the place of the royal effigy. On October 26, 1836, after an improbable journey, the Obelisk from Luxor was raised with utmost care under the watchful eyes of a crowd of Parisians who flocked to scene.

At 3300 years old, 23 meters high, and weighing 220 tons, the hieroglyph-adorned Obelisk—a gift to France and Charles X by Sultan Mehmet Ali in 1831—became one of Paris's best-known monuments. A tribute to the sun god of Ancient Egypt, the Obelisk sits in perfect alignment with the Louvre Museum and the Arc de Triomphe, installed like a flourish at the bottom of a city-sized master work.

Other works adorning the Place de la Concorde illuminate an eternal, prosperous France, with its abundance of rivers and seas. Accordingly, one of the fountains next to the Obelisk references the marine world, and the other, the Rhone and Rhine rivers, as well as the French countryside and its bounty.

Hotelphilosophy

Add: Rimini, Italy
Web:www.hotelphilosophy.net
Tel: +39 0541 775861

Hotelphilosophy is a hotel management firm that stands out in the Italian luxury hotel industry for its cosmopolitan and eclectic style that blends in with the location and takes inspiration from the most contemporary style trends. Each hotel and resort Hotelphilosophy can be described as a locus amoenus, a definition dear to the classic and Shakespearian literature that suggest a romanticized place where one can live in a private and protected atmosphere. Its approach to interior design is creating set designs for the life it wants to play a role in. Thanks to the perfect balance between contemporary taste and homage to the tradition, its group has become the benchmark for innovation and originality as well as an unforgettable hospitality experience.

Hotelphilosophy brand consist of 9 design hotels and luxury resorts set in the most extraordinary Italian locations.

Kempinski Hotel & Residences Palm Jumeirah

Add: The Crescent West, Palm Jumeirah ,P.O. Box 213208, Dubai
Tel: +971 4 444 20 00
Web: www.kempinski.com/en/dubai

The Kempinski Hotel & Residences Palm Jumeirah Dubai is brimming with exclusive perks and services. Spend the duration of your stay devoted to the finer pleasures in life, while we look after everything; travel arrangements to business services, laundry to housekeeping.Whatever you need, whether a favourite paper or chauffeur driven limousine, is just a call away. Their five-star concierge service also provides you some of the world's finest hospitality arrangements – a galaxy of choices is now yours to explore.

Ministry of Design

Add: 20 Cross Street, China Court, #03-01/06, Singapore 048422
Tel: +65 6222 5780
Web: www.modonline.com

Ministry of Design was created by Colin Seah to question, disturb and redefine the spaces, forms and experiences that surround people and give meaning to the world. An integrated spatial-design practice, MOD's explorations are created amidst a democratic "studio-like" atmosphere and progress seamlessly between form, site, object and space. They love to question where the inherent potential in contemporary design lies, and then to disturb the ways they are created or perceived – redefining the world around people in relevant and innovative ways, project by project!

This, they declare, is real change, not change for the sake of novelty. Fortified with these aspirations, they begin each distinct project anew by seeking to do 2 things – to draw deeply from the context surrounding each project, but also to dream freely so that they might transcend mere reality and convention. Each MOD project endeavours to be delightfully surprising but yet relevant, distinctly local but still globally appealing.

The response to their ethos has been overwhelming and they've received critical acclaim with a multitude of international award wins and key media coverage – these include the Rising Star of Architecture by the Monocle Singapore Survey, Gold Key Award, the International Design Awards and the President's Design Award twice over as well as feature appearances in Wallpaper, Frame and Surface. True to their multi-disciplinary profile, they've also won the Grand Prize in Saporiti Italia's design competition, and Luxury Tower was manufactured for display at the prestigious Milan Design Week 2010.

One&Only Le Saint Géran

Add: One&Only Le Saint Géran Pointe de Flacq Mauritius
Tel +230 401 1688
Email: reservations@oneandonlylesaintgeran.com
Web: www.oneandonlyresorts.com

Incomparable and distinguished, One&Only Le Saint Géran nestles peacefully in the silver white sands of its own private peninsula. Here, under the gentle sway of thousands of coconut palms towering over immaculate gardens, Mauritian influences reign supreme. Extensively remodelled in 1999, the resort's friendly calm and easy luxury is complemented by service that is attentive, yet unobtrusive. Spacious suites with terrace or balcony face out privately to the Indian Ocean.

A sheltered lagoon provides calm waters for an array of complimentary water sports ideal for adults and children alike, while the One&Only Spa offers a privileged sanctuary of quiet pampering with its own private lap pool. The lush Gary Player 9-hole golf course, located within the resort's manicured grounds, has its own clubhouse and the One&Only Golf Academy. Immortalised in Bernardin de Saint Pierre's novel 'Paul et Virginie,' the resort's peninsula is near the site of Le Saint Géran shipwreck.

One&Only Palmilla

Add: One&Only Palmilla Km 7.5 Carretera Transpeninsular San Jose Del Cabo BCS, CP 23400 Mexico
Tel +52 624 146 7000; Fax +52 624 146 7001
Email: reservations@oneandonlypalmilla.com
Web: www.oneandonlyresorts.com

One&Only Palmilla, Los Cabos Resort, a retreat of gracious splendour, where exhilaration and serenity thrive in blissful harmony. On a soft sandy beach along the idyllic Baja coast, where mighty waters of the Pacific greet playful waves in the Sea of Cortez, an ancient mountainous desert cradles a verdant oasis of sublime luxury and sheer elegance. Welcome to One&Only Palmilla, the best of Los Cabos Resorts, where exhilaration and serenity thrive in blissful harmony.

Sterling Huddleson Architecture

Add: P.O. Box 221092, Carmel, CA 93922, USA
Tel: 831-624-4363
Web: www.sterlinghuddleson.com

Sterling Huddleson Architecture is a diverse design firm dedicated to thoughtfully executed design solutions that inspire and enlighten people's personal and professional lives. Their experience includes land planning, estates design, working ranches, residential improvements, multi-housing solutions, and public environments. They enjoy working with individuals at all levels of the design and construction process, and they enjoy working on projects of all types.

The office is currently designing homes and related projects in a variety of regions including East Hampton, Coeur D Alene, Napa, Carmel, Montecito, Santa Barbara, Ojai, Los Angeles, and Arizona.

Their mission is to give their clients the homes they have always desired, with specific features tailored to the environment, and giving special attention and focus to views, light, acoustics, and other inherent qualities specific to each projects location.

The diversity of their design experience and their knowledge of construction administration is essential in creating a harmonious project team. The result is an exceptionally designed home with inspiring details and an efficient construction process.

Studio B Architects

Add: S - 260, Panchsheel Park,
New Delhi – 110017, India
Tel: +91 11 26012181
Web: http://studiobarchitect.com

Based in Delhi, the Capital of India, Studio B Architects has built a reputation for interior design of premium destination hotels and resorts in Asia, the Middle East and Africa. Their talented international team brings a diversity of experience in interior design, fresh perspective and originality, together with honed technical expertise and attention to detail results in exquisite interiors. Their high professional standards, technical resourcefulness and absolute sense of commitment to individual & corporate values finds expression in their continued preference as design consultants of choice by their valued clients for years on end.

Studio B Architects invests in great people. Their design staff is organised into groups comprising of Senior Architect, Architect and Junior Architect. Senior Architects are selected based on their superior credentials and a demonstrated ability to manage and guide a team towards producing creative and innovative design with an exceptional eye for detail. The groups rely on a network of flexible support staff that move easily between groups as projects dictate.

Everyone within the company is encouraged to challenge, recommend and invent.The work ethic to constantly strive for higher levels of creative splendor coupled with their ability to react to cultural differences gives them an obvious edge. Their design experts are a wealth of industry knowledge. Every effect is made by knowledgeable design team to help the client realize their design aspiration; the result: stunning piece of creativity.

Taleon Imperial Hotel

Add: Nevsky prospect 15, St Petersburg,
191186, Russia
Tel: +7 (812) 324-99-11 Fax: +7 (812) 324 99 57
Web: www.taleonimperialhotel.com

Space of the Taleon Imperial Hotel was made in classical style with elements of the Empire . Atlantes lobby-bar on the first floor of the hotel is made with elements of Baroque. Spa complex on the top floor is made in eclectic style with classic details of decor , fixtures and modern glass dome. Colour scheme of each room is unique. Public areas decorated in pastel colours. The following materials were used: Italian marble , beech (window frames and doors) , glass mosaic Bisazza, ceramic tile Imola Ceramica, silk wallpaper on cotton basis.

You can find the antique furniture in some guest rooms and all the furniture is made on the best factories of Europe : Ezio Belotti; Angelo Cappellini, Oak, Asnaghi, Provasi, Cabiate, Poltrona Frau (Italy), Dotzauer (Austria), Mariner (Spain), Laudarte (Italy). Floor carpets were made by the English Brintons factory.

Tuthill Architecture

Add: 701 East Broward Boulevard, Suite G, Fort
Lauderdale , Florida, 33301, USA
Tel: 954 527 0007
Email: tuthillarc@aol.com

Tuthill Architecture was established in 1979 by Robert William Tuthill and has maintained its business in Fort Lauderdale, Florida as a firm that focuses on high level design for commercial, multifamily and single family projects. The firms approach is to offer quality design with personal service through the entire design and construction process.

Wilson Associates

Add: Dallas 3811 Turtle Creek Blvd. Suite 1600 Dallas,
Texas 75219, USA
Tel: +1-214- 521-6753
Fax: +1-214-521-0207
Web: www.wilsonassociates.com

Wilson Associates was founded in 1971 by Trisha Wilson, who built the brand around creating interiors for hotels, restaurants, clubs, casinos, and high-end residential properties.

Specializing in interior architectural design, Wilson Associates is consistently named as one of the top hospitality interior design firms in the world. To date, the firm has designed and installed more than 1 million guestroom in thousands of hotels worldwide and has a client list that includes more than 20 of the world's top 100 billionaires.

The firm's design philosophy is simple: design for the market. No specific "style" or "look" is attributed to the firm. Instead, Wilson Assocates creates custom interiors for each client, conducting extensive research to fulfill each owner and operator's unique vision.

图书在版编目（CIP）数据

酒店配色与细部解析：欧式奢华 / 度本图书编著 .
-- 北京：中国林业出版社，2014.5
ISBN 978-7-5038-7520-5

Ⅰ. ①酒… Ⅱ. ①度… Ⅲ. ①饭店－建筑设计②饭店
－室内装饰设计 Ⅳ. ① TU247.4

中国版本图书馆 CIP 数据核字 (2014) 第 113620 号

中国林业出版社 · 建筑与家居出版中心

责任编辑：李丝丝

出版：中国林业出版社
（100009 北京西城区德内大街刘海胡同 7 号）
网址：http://lycb.forestry.gov.cn/
E-mail: cfphz@public.bta.net.cn
电话：（010）8322 8906
发行：中国林业出版社
印刷：北京利丰雅高长城印刷有限公司
版次：2014 年 6 月第 1 版
印次：2014 年 6 月第 1 次
开本：228mm×228mm，1/16
印张：21
字数：200 千字
定价：298.00 元